Mechanics of Robotic Manipulation

Intelligent Robots and Autonomous Agents
Ronald C. Arkin, editor

Behavior-Based Robotics, Ronald C. Arkin, 1998

Robot Shaping: An Experiment in Behavior Engineering, Marco Dorigo and Marco Colombetti, 1998

Layered Learning in Multiagent Systems: A Winning Approach to Robotic Soccer, Peter Stone, 2000

Evolutionary Robotics: The Biology, Intelligence, and Technology of Self-Organizing Machines, Stefano Nolfi and Dario Floreano, 2000

Reasoning about Rational Agents, Michael Wooldridge, 2000

Introduction to AI Robotics, Robin R. Murphy, 2000

Strategic Negotiation in Multiagent Environments, Sarit Kraus, 2001

Mechanics of Robotic Manipulation, Matthew T. Mason, 2001

Mechanics of Robotic Manipulation

Matthew T. Mason

A Bradford Book
The MIT Press
Cambridge, Massachusetts
London, England

©2001 Massachusetts Institute of Technology

All rights reserved. No part of this book may be reproduced in any form by any electronic or mechanical means (including photocopying, recording, or information storage and retrieval) without permission in writing from the publisher.

This book was set in Times-Roman by the author and was printed and bound in the United States of America.

Library of Congress Cataloging-in-Publication Data

Mason, Matthew T.
  Mechanics of robotic manipulation / Matthew T. Mason.
    p. cm.—(Intelligent robotics and autonomous agents)
  "A Bradford book."
  Includes bibliographical references and index.
  ISBN 0-262-13396-2 (hc. : alk. paper)
  1. Manipulators (Mechanism). 2. Robotics. I. Title. II. Series.

TJ211 .M345   2001
629.8'92—dc21   2001030226

Contents

Preface   ix

Chapter 1   Manipulation   1
1.1   Case 1: Manipulation by a human   1
1.2   Case 2: An automated assembly system   3
1.3   Issues in manipulation   5
1.4   A taxonomy of manipulation techniques   7
1.5   Bibliographic notes   8
        Exercises   8

Chapter 2   Kinematics   11
2.1   Preliminaries   11
2.2   Planar kinematics   15
2.3   Spherical kinematics   20
2.4   Spatial kinematics   22
2.5   Kinematic constraint   25
2.6   Kinematic mechanisms   34
2.7   Bibliographic notes   36
        Exercises   37

Chapter 3   Kinematic Representation   41
3.1   Representation of spatial rotations   41
3.2   Representation of spatial displacements   58
3.3   Kinematic constraints   68
3.4   Bibliographic notes   72
        Exercises   72

Chapter 4   Kinematic Manipulation   77
4.1   Path planning   77
4.2   Path planning for nonholonomic systems   84
4.3   Kinematic models of contact   86
4.4   Bibliographic notes   88
        Exercises   88

Chapter 5 Rigid Body Statics 93
5.1 Forces acting on rigid bodies 93
5.2 Polyhedral convex cones 99
5.3 Contact wrenches and wrench cones 102
5.4 Cones in velocity twist space 104
5.5 The oriented plane 105
5.6 Instantaneous centers and Reuleaux's method 109
5.7 Line of force; moment labeling 110
5.8 Force dual 112
5.9 Summary 117
5.10 Bibliographic notes 117
    Exercises 118

Chapter 6 Friction 121
6.1 Coulomb's Law 121
6.2 Single degree-of-freedom problems 123
6.3 Planar single contact problems 126
6.4 Graphical representation of friction cones 127
6.5 Static equilibrium problems 128
6.6 Planar sliding 130
6.7 Bibliographic notes 139
    Exercises 139

Chapter 7 Quasistatic Manipulation 143
7.1 Grasping and fixturing 143
7.2 Pushing 147
7.3 Stable pushing 153
7.4 Parts orienting 162
7.5 Assembly 168
7.6 Bibliographic notes 173
    Exercises 175

Contents

Chapter 8   Dynamics   181

8.1   Newton's laws   181
8.2   A particle in three dimensions   181
8.3   Moment of force; moment of momentum   183
8.4   Dynamics of a system of particles   184
8.5   Rigid body dynamics   186
8.6   The angular inertia matrix   189
8.7   Motion of a freely rotating body   195
8.8   Planar single contact problems   197
8.9   Graphical methods for the plane   203
8.10 Planar multiple-contact problems   205
8.11 Bibliographic notes   207
      Exercises   208

Chapter 9   Impact   211

9.1   A particle   211
9.2   Rigid body impact   217
9.3   Bibliographic notes   223
      Exercises   223

Chapter 10   Dynamic Manipulation   225

10.1 Quasidynamic manipulation   225
10.2 Briefly dynamic manipulation   229
10.3 Continuously dynamic manipulation   230
10.4 Bibliographic notes   232
      Exercises   235

Appendix A   Infinity   237

References   241
Author Index   247
Subject Index   249

Preface

This book is written for anyone mystified and intrigued by manipulation. In its most general form, "manipulation" refers to a variety of physical changes to the world around us: moving an object; joining two or more objects by welding, gluing, or fastening; reshaping an object by cutting, grinding, or bending; and many other processes. However, this book, like the vast bulk of manipulation research, addresses only the first of these forms of manipulation: moving objects. Even with this restriction, we still have a number of different processes to consider: grasping, carrying, pushing, dropping, throwing, striking, and so on.

Likewise, we will address only robotic manipulation, neglecting manipulation by humans and other animals, except for inspiration and occasional philosophical points. But "robotic" manipulation should not be construed too narrowly—perhaps "machine manipulation" would be a better phrase. We will include any form of manipulation by machines, from doorstops to automated factories.

The book draws on material from two fields: classical mechanics and classical planning. Much of the book is devoted to classical mechanics, as it applies to manipulation processes. The goal of understanding manipulation provides an unusual perspective on classical mechanics, and will force us to address some peculiarities not treated in most texts.

The second component of the book is classical planning. By this I mean state-space approaches, where an explicit model of the possible actions allows the planner to search through various sequences until a satisfactory one is found. Two difficulties have to be resolved. First, classical mechanics typically gives multi-dimensional continuous state spaces, rather than the discrete state spaces that are better suited to search. Second, the robot does not generally have perfect information, and may not know the actual state of the task. Sometimes a plan has to address this gap between the task state and the robot's estimate of task state. Both of these factors—high-dimensional continuous state spaces, and uncertainty—add to the complexity of manipulation planning.

This book differs from most previous books in its focus on *manipulation* rather than *manipulators*. This focus on processes, rather than devices, allows a more fundamental approach, leading to results that apply to a broad range of devices, not just robotic arms. The real question of manipulation is how to move objects around, not how to move an arm around. Humans seem to solve this problem by exploiting all the available resources, using any convenient surface to align objects, tapping and shaking things that are inconvenient to grasp, using any convenient object as a tool to poke and push, and so forth. This ability is best observed when a human uses his own hands, but it is also apparent when a human programs a robotic manipulator. Any faithful attempt to explain manipulation must address a broad range of manipulation techniques.

In robotics, any theory that reaches a certain stage of maturity can be tested. If a theory is complete enough, and constructive, we can build a robot that embodies the theory, and then test the validity and scope of the theory through experiments with the robot. It is relatively straightforward, in principle, to build a robot combining classical mechanics and classical planning. We construct a robot that has a computational model of the task, including shapes and other relevant physical parameters of the objects in the scene. Using classical mechanics, the robot is also able to predict the effect of any actions it might wish to consider. Now if the robot is given some goal, it can simulate various sequences of actions, searching for a plan that will achieve the goal.

Such a robot is the ultimate rationalist—its faith in Newtonian (or Aristotelian or whatever) mechanics is absolute, and its actions are derived from first principles to satisfy its goals. It is a near-perfect marriage of theory to experiment. To address theoretical concerns, we can design the robot from formal models of mechanics and search algorithms, obtaining a formal entity susceptible to theoretical investigation. We can prove theorems about its performance, and we have explicit formal hypotheses that characterize its validity. To address experimental concerns, we can implement the design, obtaining a physical system that can be tested experimentally. When theory and experiment agree, we have evidence for the validity of the theory and the fidelity of its implementation. When they disagree, we have hints about appropriate changes to the theory or its implementation.

Perhaps even more important is the value that this approach places on developing an effective, constructive theory. There is sometimes a big difference between a theory that *should* work and a theory that *does* work. The process of closing that gap is an important force in driving the field to address the important problems.

What kind of theory should we be attempting to build? Will there be a neat solution—some simple idea that will allow us to build robots with human-like abilities? Comparison with related engineering disciplines suggests otherwise. Nobody expects a single neat theory for the problem of building a car or a rocket. Artifacts of great complexity depend on a large diverse collection of scientific and engineering results, and a robot competitive with a human being will surely be more complex than anything we have constructed before. This book does not propose to describe the solution, or even an outline of the solution. Rather, it attempts to outline one specific line of scientific inquiry, which we may hope addresses one of the central problems of robotic manipulation.

This book began as course notes for a graduate course, "Mechanics of Manipulation," which is part of the curriculum of the Robotics PhD program at Carnegie Mellon. The students had varied backgrounds, but most had an undergraduate degree in engineering, science, or mathematics. Occasionally an advanced undergraduate has taken the course; most have done well. Most students, but not all, have already taken a course in manipulator kinematics, dynamics, and control. Term projects have been an integral part of the course. Each student, or in some cases a small team, is expected to choose and explore a manipulation problem. Some of the possibilities are: building a card house, cracking a whip,

Preface

throwing a frisbee, building a stack of dominos with maximum overhang, snapping one's fingers, throwing a top, working a yoyo, solving the ball-in-cup game, various types of juggling, and so on. A typical project might analyze some simplified version of the problem, produce a simple planning system, or focus on some well-defined aspect of the process. Papers on many such problems may be found in the back issues of *Scientific American* and the *American Journal of Physics*. Some of the projects address more respectable manipulation problems, which can be found in *International Journal of Robotics Research*, or *Proceedings of the IEEE International Conference on Robotics and Automation*, among others.

For the benefit of those teaching from this book or doing the problems, the figures will be posted on my web page at http://www.cs.cmu.edu/~mason, possibly with additional notes on teaching, or solutions for the problems.

I would like to thank my advisors and colleagues: Berthold Horn, Tomas Lozano-Pérez, Marc Raibert, Mike Erdmann, Randy Brost, Yu Wang, Ken Goldberg, Alan Christiansen, Kevin Lynch, Srinivas Akella, Wes Huang, Garth Zeglin, Devin Balkcom, Siddh Srinivasa, John Hollerbach, Russ Taylor, Ken Salisbury, Dan Koditschek, Bruce Donald, Illah Nourbakhsh, Ben Brown, Tom Mitchell, Dinesh Pai, Al Rizzi, Takeo Kanade, and Allen Newell.

Several people have read drafts, adopted the book for classes, or provided assistance and encouragement in other ways: Anil Rao, Howard Moraff, Carl Harris, Charlie Smith, Ian Walker, Mike McCarthy, Zexiang Li, Richard Voyles, Yan-Bin Jia, Terry Fong, Kristie Seymore, Elisha Sacks, and the students of 16-741. Jean Harpley and Mark Moll helped with the final preparation of the manuscript, and Sean McBride drew the drawings in chapter 1.

I would like to thank Mary, Timm, and Kate for their support and sacrifices.

I would like to thank the National Science Foundation for support under grants IRI-9114208, IRI-9318496, IIS-9900322, and IIS-0082339.

I would like to thank everyone whose ideas I have borrowed. In some cases the contributions have been much greater than might be appreciated from scanning the bibliographies or the index.

1 Manipulation

Manipulation is the process of using one's hands to rearrange one's environment. There are many facets to manipulation. Manipulation is an art, since it is practiced by all of us, without any systematic or fundamental understanding of the process. Manipulation is also an engineering discipline, for there are some systematic tools for applying robotic manipulation to various problems. Finally, manipulation is also a science, since it is a process that engages our curiosity, which we can explore using scientific methods.

Manipulation is accomplished in many different ways. To begin the chapter, we will consider two examples of manipulation systems. Thus we establish the subject of the study—the set of phenomena to be explored and explained. The rest of the chapter catalogs this diverse set of manipulation techniques, in terms of the underlying mechanics. The chapter ends with an outline of the book.

1.1 Case 1: Manipulation by a human

Our first example has the form of a thought experiment, but it is a familiar scenario drawn from everyday life, one which can easily be re-enacted if desired (figure 1.1). Consider the processes by which the dealer in a card game gathers the cards, assembles a neat pack, shuffles the pack, deals the cards, collects his own hand, and arranges it. Although it is hard to analyze the process with any precision, even a superficial analysis is revealing. First the dealer sweeps the cards together into a heap, then he squeezes the heap and taps it on the table until a neat pack is obtained. Shuffling is most commonly accomplished by splitting the pack into two halves, bending them, then releasing them in a sequence so that the two halves are interleaved. Then the pack is neatened again by squeezing and tapping.

Now the left hand is fashioned into a mechanism that presents isolated cards in succession, which the right hand grasps and throws. The throw includes a spin that stabilizes the card's orientation.

Now all the players gather and neaten their hands. They arrange the cards by grasping and re-inserting subsets of the cards. Again the hand is neatened by squeezing, then carefully spread by a technique involving controlled slip among the cards.

Several characteristics of the process are intriguing. Handling of individual cards is kept to a bare minimum. Some of the cards are not handled individually at all, except for the throw. Not once is an isolated card brought from one rest position to another by the technique of being rigidly attached to the hand and arm. Instead the dealer uses sweeping, tapping, squeezing, throwing, controlled slip, and some additional techniques that defy simple descriptions.

Figure 1.1
An example of skilled human manipulation: gathering, straightening, shuffling, and dealing cards.

Manipulation

The process requires a fair degree of skill. Young children are much less adept, and learning the skill takes time and practice.

The techniques are sensitive to the characteristics of the cards and the table. New cards are awkward because they are too slippery and too stiff. Dirty cards are too sticky. The cards need to be precisely identical in dimensions, and the right stiffness. A pack of cards cut by hand from ordinary writing paper requires very different methods.

In some respects the operations are strongly sensor-driven, as when turning cards that are out of alignment with the rest of the pack, or gathering cards that are scattered around the table. At times, though, sensors play a minimal role. For example, the final neatening of the pack does not rely on vision to identify the misaligned cards. Rather, squeezing and tapping the deck takes care of the cards without the necessity of identifying them.

Two things stand out very strongly. One is the efficiency and skill displayed. The other is the adaptability. Although a change in the apparatus may drastically affect efficiency, the human can adapt, falling back on more conservative techniques.

1.2 Case 2: An automated assembly system

Our second example is the automated assembly line illustrated in figure 1.2, integrating industrial robotic manipulators and a variety of equipment for transporting, orienting, and presenting parts to the robots. We will base the example on the Sony SMART system, but the basic elements are common to many industrial systems.

The problem is to assemble a small consumer electro-mechanical product, such as a tape recorder or a camera. The assembly takes place on a work fixture that is designed to hold the device and carry it accurately from one station to the next. At each station a robot performs a number of operations. In order to minimize the number of stations, each robot has up to six different end-effectors, all of which are mounted on a turret head. Thus each robot can perform six or more different steps, by selecting the relevant effectors as appropriate.

Parts are presented to the robots in pallets. Each pallet has an array of nests, each nest holding a single part, in an attitude appropriate for the robot to grasp the part. These pallets in turn are transported on a second conveyor system, which can retrieve a pallet from storage on demand, and transport the pallet to the robot that requested it.

The pallets are filled with parts by an APOS machine. The pallet is tipped at a slight angle, and parts are shoveled onto it. A vibratory motion of the pallet allows the parts to slide down the plate and off into an overflow bin, but some of the parts fall into the nests. The nest shape and vibratory motion are designed so that parts will come to rest only in

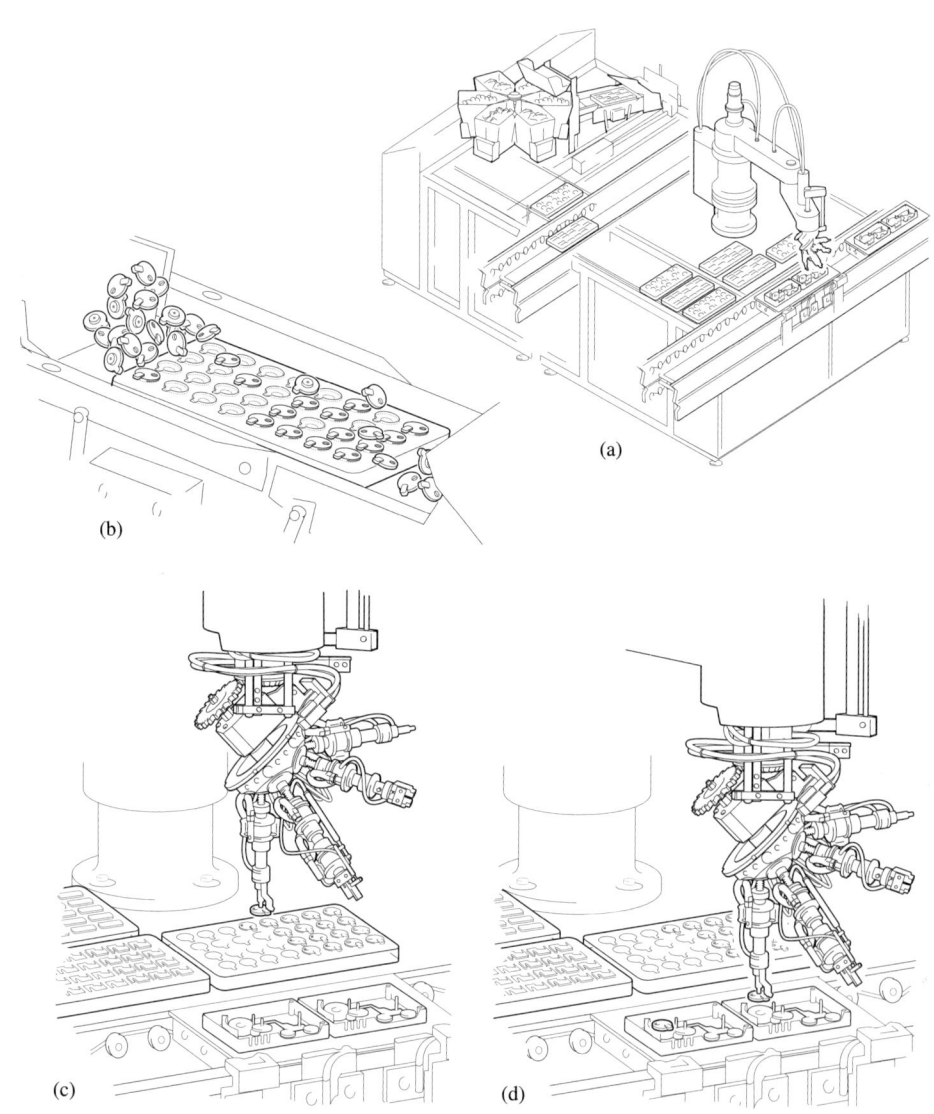

Figure 1.2
The Sony SMART system. (a) System layout showing work conveyor, parts pallets, robot with turret tool head, pallet conveyor, and APOS parts orienting system. (b) Closeup of APOS filling pallet with oriented parts. (c) Picking a part. (d) Assembling a part into the product.

Manipulation 5

the desired orientation. After some pre-determined time, the pallet is filled with oriented parts. It is then unloaded from the APOS machine and delivered to a robot or to storage.

The design of the end product is optimized to simplify the assembly process (*design for assembly*). In particular, the products are designed to be assembled one part at a time, with almost all assembly motions being vertical. Parts are designed to be easy to handle. For instance, a single complex plastic shape might replace several simpler shapes, with springs replaced by flexible elements of the part. The parts are also designed to reduce the difficulty of feeding and orienting the part, and to reduce the difficulty of assembly, so that mating features actually help guide the assembly into place.

Similarly, the end-effectors and the nests are designed to speed the process. In fact, some reflection reveals that assembly is ubiquitous in this process. Each part is oriented by assembling it with a pallet nest. Each part is grasped by assembling the robot gripper with the part. Assembly of the part with the work in progress is actually the third assembly step in this process.

Shape interactions are the dominant phenomenon in this system. The critical steps are: (1) part locating by the interaction of a part with a nest in the APOS system; (2) part grasping by mating a special-purpose effector with a feature of the part; and (3) part placement, which might involve interactions between the part and the work-in-progress. Sensors play a small but important role, allowing the robot to detect a grasp failure and proceed by going on to the next part in the pallet. The manipulator is used in the simplest possible way, as a pick and place device.

1.3 Issues in manipulation

There are striking differences between the human card player and the automated factory. Humans have thousands of sensors and actuators, and the intelligence to coordinate them and adapt them to the task at hand. A single robot has only a few sensors and actuators, and lacks the intelligence to adapt them to a new task without human help. Even if we consider the factory as a whole, with hundreds of sensors and actuators, they are still configured and programmed for a single task, or a closely related set of tasks.

But we can see through these differences, by focusing on the decisions made by the two systems. Some of these decisions are online: decisions made and quickly acted upon, perhaps using information just acquired. Some of these decisions are offline: techniques that are learned through practice and can be applied without being reinvented. Some might be considered off-offline: decisions that were made at design time, such as how many fingers to have, how to configure the sensors, and so forth.

When we focus on decisions, we see that the main difference between the systems is the human's much greater capacity for making decisions online. The robotic system makes very few online decisions. The only variations in its actions occur as a result of error conditions, or depend upon the arrival of a particular pallet. All other decisions were made offline by the system designers as the system was being designed and programmed. For the human, it is harder to determine whether a decision is made online or offline. It seems clear that the human makes many more online decisions than the factory system. But it is also clear that the human system still involves many offline decisions. At the least, to develop good skill requires long practice, comprising decisions occuring over a long period of time. And of course the more basic aspects of the human system were determined as the species evolved.

Although the *time* of the decisions varies greatly between humans and robots, the decisions can otherwise be quite similar: how to configure sensors, actuators, and mechanical structures; how to organize and coordinate sensory information with actuator innervation; what patterns of shape, motion, and force will produce a desired result. A theory of manipulation ought to provide a means for making these decisions, for solving manipulation problems, whether offline or online.

What are the manipulation problems that have to be solved? Some characteristics of manipulation systems are dictated by the inherent properties of the problem, and are shared by all solutions to the problem. This observation can serve as a guide to our study, helping us to focus on phenomena and approaches that are fundamental to manipulation, transcending the details of any particular technology. The card playing system and the robotic assembly system raise some fairly basic problems, which we can take as representative of manipulation in general. For example,

- Show that an object, or set of objects, is in a stable configuration. As an example, every stage in the assembly must be stable, while the robot puts in the next piece.
- Show that an object is *not* in a stable configuration. A strategy for aligning cards can be designed by finding a set of finger motions, such that no out-of-place card is stable. Similarly an APOS orienting-nest design should have the property that no out-of-place part is stable.
- Given a fixed object, a mobile object, and an applied force, show that the mobile object converges locally to a particular location relative to the fixed object. For example, this approach can be used to analyze a gripper's ability to grasp a particular object in a predictable way.
- Given a fixed object and a mobile object, design a vibrational motion so that the mobile object converges *globally* to a particular location relative to the fixed object.
- Construct a throwing motion to deliver an object accurately, and minimize the energy of the object subsequent to impact. This may be a good way to synthesize a card-throwing motion, with the additional constraint that the card stay face down.

Problems of this kind fall between mechanics and planning, and can be phrased either as problems of analysis or as problems of synthesis. Phrased in analytic form, we have a mechanics problem. Unfortunately, for many of these problems there is not a general solution, especially when we include limitations on the information available for the solution. Phrased in synthetic form, we have a planning problem. This allows us to make choices that restrict the scope of the problem, possibly allowing a solution.

We will take as the over-riding problem: how does a robot transform goals into actions? To narrow the focus a bit, we will assume that a robot is given some goals, generally requiring some re-arrangement of the objects surrounding it. We will not worry about higher-level issues, such as how a robot transforms a higher-level goal, such as profit, to lower-level goals, such as grasping a wrench.

1.4 A taxonomy of manipulation techniques

We will focus on one approach, which might be called *analytical manipulation*. To decide what to do, the robot will use an analytical model of the task mechanics. Before trying an action, the robot can use its model to predict the consequences of the action. Rather than classifying robots according to their physical structure, or their computational architecture, we will classify robots according to their models of task mechanics. Thus we distinguish a *Newtonian robot*, which derives its actions from Newton's laws, from an *Aristotelian robot*, which thinks things move only when something pushes them. We could also have an *empirical robot*, which uses a model built up from observations, rather than one based on an axiomatic system.

All of the robots we will explore here are variants of the Newtonian robot, although some of them could also viewed as Aristotelian robots. All of them make decisions based on the familiar techniques of classical mechanics. Classical mechanics is usually presented in a sequence that begins with kinematics, then moves to statics, and finally to dynamics. We can employ the same progression to construct a hierarchy of manipulation techniques:

- Kinematic manipulation. An action or sequence of actions that can be derived from kinematic considerations alone. For example, if the task specification is a particular motion of the end-effector, the motions of the arm mechanism can be determined kinematically.

- Static manipulation. An action or sequence of actions that can be derived from statics and kinematics. For example, to place an object on a table, it is necessary to determine a stable rest attitude for the object (statics) and to produce the motions necessary to move the object there (kinematics).

- Quasistatic manipulation. In manipulation tasks, inertial forces are often dominated by frictional forces and impact forces. Analysis that neglects inertial forces is often called quasistatic analysis. For example, when straightening a pack of cards by squeezing it, the inertial forces of the cards are negligible.
- Dynamic manipulation. Finally, when inertial forces are an important part of the process, we have dynamic manipulation. For example, throwing a card so that it lands properly and never flips over, depends on the inertial properties of the card.

Some care must be exercised in applying this taxonomy. It refers to how actions are *derived*, rather than referring to the actions themselves. And in most instances, we do not know how the actions were derived, since the derivation occurred inside a human's head. The classification is actually a subjective one, depending on the observer's own model of manipulation.

Our chief use of the taxonomy is to organize this book. We can follow the traditional progression from kinematics to dynamics, exploring mechanics with the goal of understanding manipulation processes. The book begins with kinematics (chapters 2 and 3), followed by a chapter on kinematic manipulation. The chapters on statics and friction are followed by a chapter on quasistatic manipulation. And the chapters on dynamics and rigid-body impact are followed by a chapter on dynamic manipulation.

1.5 Bibliographic notes

For a broad understanding of the role of manipulation and the human hand, (Bronowski, 1976) is essential. Also highly recommended are (Napier, 1993) and (Wilson, 1998).

Although humans seem to be the best, many animals are impressive manipulators. Savage-Rumbaugh and Lewin (1994) provide several interesting comparisons of apes and humans. Collias and Collias (1984) provide a very rewarding description of bird nest building.

Fujimori (1990) describes the Sony SMART system. The taxonomy of manipulation is first described by Kevin Lynch and myself (1993). For a broader treatment of automated assembly, see (Boothroyd, 1992).

Exercises

Exercise 1.1: The taxonomy of manipulation techniques refers to the robot's model of task mechanics. Does a robot *have* to have a model of the task mechanics? Does an ant have a model of the task mechanics? Construct arguments on both sides of the question.

Exercise 1.2: Suppose you are given a robot and you want to know whether it uses "analytical manipulation" or not. You are allowed to perform any experiments, to take it apart, etc. How could you tell whether the robot was analytical or not?

Exercise 1.3: Do you use "analytical manipulation" or "empirical manipulation"? Or both, or neither? Perhaps sometimes one, sometimes the other? Consider the processes of evolution, how experience might be encoded and accessed in the brain, and how to generalize from past experience to deal with a current problem.

Exercise 1.4: Analyze some manipulation task such as playing cards. Identify the different stages, and for each stage describe the mechanical processes, the control and decision processes, the information required to perform the actions, and the source of information.

2 Kinematics

Kinematics means the study of motion, without regard for the cause of the motion. There are many reasons for us to begin with kinematics. First, manipulation is typically aimed at moving objects around, so the principles of kinematics are relevant to almost any manipulation process. Second, many manipulation processes are entirely kinematic in nature. These processes were termed *kinematic manipulation* in chapter 1, and are addressed in chapter 4. Finally, kinematics also is the traditional first step in treatments of classical mechanics, and it can play the same role here.

This chapter's treatment of kinematics differs somewhat from more general treatments, owing to our goal of understanding the processes of manipulation. This chapter presents the basic principles of rigid body motion, but simultaneously illustrates and motivates these principles by applying them to manipulation. You will already be familiar with many of the concepts in this chapter, such as *translation* and *rotation*. Our goal is to augment your working knowledge with a more precise foundation, but we do not strive for complete rigor, which might unnecessarily obscure the important concepts.

2.1 Preliminaries

Before focusing on rigid bodies, we digress briefly to consider more general systems. We will consider a *system* to be a set of points in some *ambient* space: two-dimensional Euclidean space ($\mathbf{E}^2$) for planar kinematics, three-dimensional Euclidean space ($\mathbf{E}^3$) for spatial kinematics, or the surface of a sphere ($\mathbf{S}^2$) for spherical kinematics. In the general case, each point might be capable of independent motion. For the general case, then, let $\mathbf{X}$ be the ambient space—$\mathbf{E}^2$, $\mathbf{E}^3$, or $\mathbf{S}^2$.

DEFINITION 2.1: A **system** is a set of points in the space $\mathbf{X}$.

DEFINITION 2.2: A **configuration** of a system is the location of every point in the system.

DEFINITION 2.3: **Configuration space** is a metric space comprising all configurations of a given system.

How do we know that a metric exists for the space of all configurations? We could define a metric on the configuration space for any system, by choosing a few points $\{x_i\}$ in the system, and defining the distance between two configurations $d(D_1, D_2)$ to be the max-

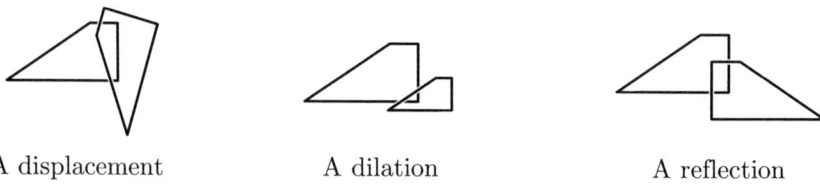

A displacement   A dilation   A reflection

Figure 2.1
Transformations, rigid and otherwise.

imum distance between corresponding points, $d(D_1, D_2) = \max\{d(D_1(x_i)), d(D_2(x_i))\}$. Note, however, that this metric involves an arbitrary choice of the x_i. In fact, for many of the configuration spaces we will consider, rigid body configuration space in particular, every metric involves arbitrary choices.

DEFINITION 2.4:   The **degrees of freedom** of a system is the dimension of the configuration space. (A less precise but roughly equivalent definition: the minimum number of real numbers required to specify a configuration.)

Now we can move on to the concept of rigid bodies and rigid motions.

DEFINITION 2.5:   A **displacement** is a change of configuration that does not change the distance between any pair of points, nor does it change the handedness of the system.

(I use the term "handedness" in favor of the usual term "orientation".)

DEFINITION 2.6:   A **rigid body** is a system that is capable of displacements only.

A transformation that uniformly changes the size of a body is called a scale transformation, or a dilation. A transformation that changes the handedness of a body is called a reflection.

System	Configuration	DOFs
point in plane	x, y	2
point in space	x, y, z	3
rigid body in plane	x, y, θ	3
rigid body in space	$x, y, z, \phi, \theta, \psi$	6

Kinematics

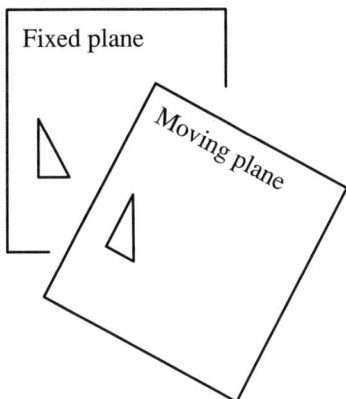

Figure 2.2
Moving and fixed planes.

It is convenient to consider a displacement to apply to every point in the space, not just the points in some rigid body. As an example, consider a triangle capable of rigid motion in the plane (figure 2.2). We can imagine that the triangle is drawn on a *moving plane*, and we can refer to the base plane as the *fixed plane*. Any motion of the triangle determines a motion of the entire moving plane. Thus we can talk about the motion of an arbitrary point, whether or not this point is actually part of the body. Also we can study displacements of the plane, as a means of studying displacements of any rigid body in the plane.

DEFINITION 2.7:   A **rotation** is a displacement that leaves at least one point fixed.

It is important to note a distinction between two similar concepts: rotation about some given point such as the origin or the center of mass; versus rotation about an unspecified point. For a general rotation, the fixed point may be anywhere in the space.

DEFINITION 2.8:   A **translation** is a displacement for which all points move equal distances along parallel lines.

At times it is convenient to use algebraic concepts and notation in discussing displacements. Every displacement D can be described as an operator on the underlying space, mapping every point x to some new point $D(x) = x'$. The product of two displacements is the composition of the corresponding operators, i.e. $(D_2 \circ D_1)(\cdot) = D_2(D_1(\cdot))$. The inverse of a displacement is just the operator that maps every point back to its original position.

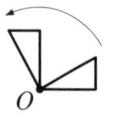  

Rotation about O | Rotation about a point on the body | Rotation about a point not on the body

Figure 2.3
Various rotation centers.

The identity is the null displacement, which maps every point to itself. These observations can be summarized as follows:

THEOREM 2.1: The displacements, with functional composition, form a group.

These groups have names. For the Euclidean spaces we have the *special Euclidean groups* **SE**(2) and **SE**(3). For the sphere we have the *special orthogonal group* **SO**(3). "Special" means that they preserve handedness. "Orthogonal" refers to the connection to orthogonal matrices, which will be covered in chapter 3.

The next question is "Do displacements commute?" That is, does $D_2(D_1(\cdot)) = D_1(D_2(\cdot))$ hold for arbitrary displacements D_1 and D_2? The answer is no. Figure 2.4 gives a counter-example—two spatial rotations that give different results when taken in different orders.

There are many different ways to describe a displacement, but the most familiar is to decompose the displacement into a rotation followed by a translation.

THEOREM 2.2: For any displacement D of the Euclidean spaces $\mathbf{E}^2$ or $\mathbf{E}^3$, and any point O, D can be expressed as the composition of a translation with a rotation about O.

Proof Let O' be the image of O under D. Let T be the translation taking O to O', and let T^{-1} be its inverse. Then the displacement $T^{-1} \circ D$ leaves O fixed, so it is a rotation; call it R. Then $T \circ R = T \circ T^{-1} \circ D = D$ is the desired decomposition. Alternatively, we could define a rotation $S = D \circ T^{-1}$. So there two ways of decomposing D into a rotation and a translation: $T \circ R$ or $S \circ T$. ∎

For many purposes the decomposition into a rotation followed by a translation is a good way to describe displacements, but take note that it is not a canonical description— the decomposition depends on the reference point O.

Kinematics

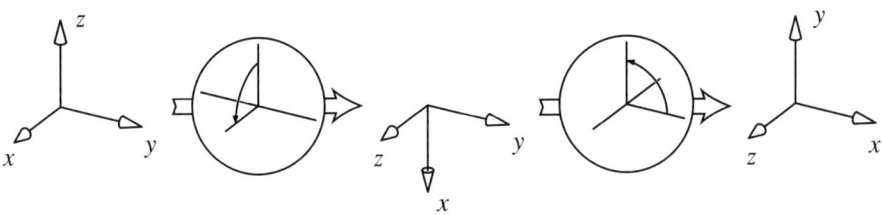

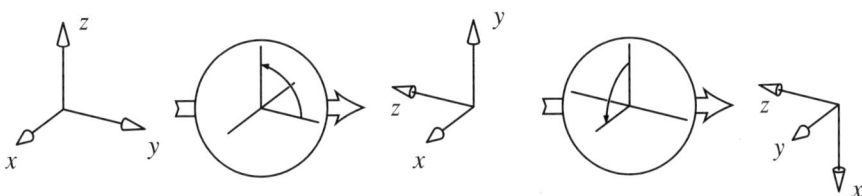

Figure 2.4
Spatial rotations do not generally commute.

COROLLARY 2.1: Given any point O, any differential motion or velocity can be decomposed into a translational part and a rotational part about O.

2.2 Planar kinematics

To explore basic kinematics further, we must consider the underlying ambient spaces separately. This section focuses on *planar kinematics*. The main topic is how the special cases of rotation and translation relate to the general displacement. Theorem 2.2 showed that any displacement can be described as a translation followed by a rotation. But for planar kinematics we can go much further—every displacement can be described as *either* a translation *or* a rotation. In fact, if we allow the small mathematical convenience of allowing points at infinity, translations can be treated as rotations, so that every planar displacement can be viewed as a rotation. First we must lay out some of the basic properties of displacements, rotations, and translations in the plane.

THEOREM 2.3: A planar displacement is completely determined by the motion of any two points.

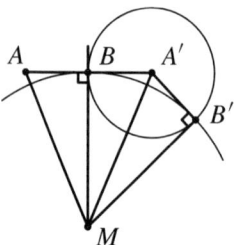

Figure 2.5
Construction for proof of theorem 2.4.

Proof The displacement is completely determined if the motion of every point in the plane is determined. Given the motion of two points, let the origin be one point, and choose the x axis to point at the second point. Choose the y axis to give a right-handed coordinate frame. The motion of the two points determines the motion of the coordinate frame. Given the coordinates of any other point P in the plane, we can use the coordinate frame to construct the image P'. ∎

Recall that a displacement can be decomposed into the product of a rotation and a translation (theorem 2.2). Unfortunately, this decomposition depends on the choice of a reference point, so it is not a canonical description. The next result gives a means of describing planar displacements that is often superior.

THEOREM 2.4: Every planar displacement is either a translation or a rotation.

Proof Let D be an arbitrary planar displacement, let A be an arbitrary point in the plane, and let A' be its image under D. If $A = A'$ then D is by definition a rotation, so we henceforth assume that A' is distinct from A. Let B be the midpoint of the line segment $\overline{AA'}$, and let B' be the image of B.

If B' is collinear with A, A', and B, then there are only two choices for B' that preserve distance from A'. One choice leaves B fixed, giving a rotation. The other choice gives a translation by the vector $\overrightarrow{AA'}$.

The only case left is that B maps to a distinct B' not on the line through A, B, and A', which is illustrated in figure 2.5. Construct a perpendicular to AB at B, and a perpendicular to $A'B'$ at B'. These perpendiculars are not parallel, since AB and $A'B'$ are not parallel. Let M be the intersection of the perpendiculars. We will prove M is fixed. Consider the displacement mapping A to A' and M to itself. Where does that displacement map B? Since it must preserve distance from A and from M, we identify two candidates by intersecting

Kinematics

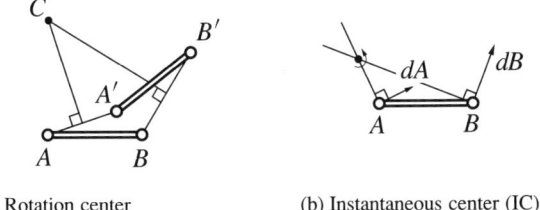

(a) Rotation center   (b) Instantaneous center (IC)

Figure 2.6
Generic cases of constructing a rotation center and an instantaneous center.

circles. One candidate, B itself, is excluded because it would correspond to the identity. The other candidate is B'. This proves that our hypothesized rotation about M is the original displacement D. ∎

The fixed point of a planar rotation is easily constructed from the motions of two points A and B. We construct the perpendicular bisectors of $\overline{AA'}$ and $\overline{BB'}$, and intersect. Figure 2.6(a) shows the generic case, where the perpendicular bisectors have a single point of intersection. Two other possibilities must be considered. If the two perpendicular bisectors are identical, we must repeat the construction, after substituting for A or B any point not collinear with A and B. The last case is a translation, for which the two perpendiculars are parallel. We can treat this case as a rotation of zero degrees about a point at infinity. (A brief treatment of points at infinity is given in appendix A.) Hence we restate theorem 2.4 as:

Every planar displacement is a rotation about a point in the projective plane.

COROLLARY 2.2: Generalizing to differential motions, every plane velocity is an angular velocity about a center in the projective plane. To construct the point, we take the perpendiculars to dA and dB and intersect.

These results have given rise to a number of terms, with no firm conventions on their use. *Rotation center* and *rotation pole* refer to finite rotations; *instantaneous center (IC)*, *velocity center,* and *velocity pole* refer to velocities.

Centrodes

Until now, we have focused on single displacements. Now we consider continuous motions, as for an object whose configuration is changing as a continuous function of time. When a

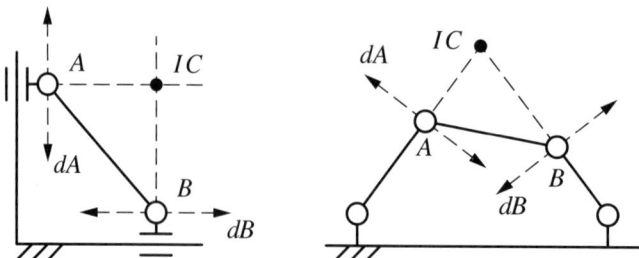

Figure 2.7
Construction of ICs for two four-bar linkages.

motion is parameterized by time, we use the term *trajectory*, described by a curve $q(t)$ in configuration space. In this section we will ignore the time element, and consider the *path* of the motion, described by a curve $q(s)$ in configuration space, whose parameterization may be considered arbitrary.

The main result is that any planar motion path can be described by two curves in the plane, called *centrodes*. One curve, the moving centrode, rolls without slipping along the second curve, called the fixed centrode. The centrodes provide a canonical description of planar motions, which has the additional advantage of being easily understood.

The fixed centrode is the locus of the rotation pole in the fixed plane. The moving centrode is the locus of the rotation pole in the moving plane. It should be evident that as the moving centrode rolls without slipping on the fixed centrode, the instantaneous center is at the contact between the two curves.

The main difficulty in constructing centrodes by hand is to plot the centers in the moving plane. The easiest way is to construct the fixed centrode first, then use a transparent sheet of plastic for the moving plane, plotting the centers on the moving plane as it reproduces the motion in question.

We will illustrate the approach using planar four-bar linkages (figure 2.7). Each four-bar linkage consists of a motionless base link, two links whose motion is a rotation about a fixed point or a translation along a fixed axis, and a fourth link, the *coupler link*, which makes a fairly complicated motion. At any given instant instant, we can describe the constraints on the coupler link by noting that the motions of points A and B are constrained. The rotation center of the coupler link must be on a line perpendicular to dA and also on a line perpendicular to dB. This completely determines the instantaneous center of the coupler link unless the linkage is passing through some degenerate configuration.

Kinematics

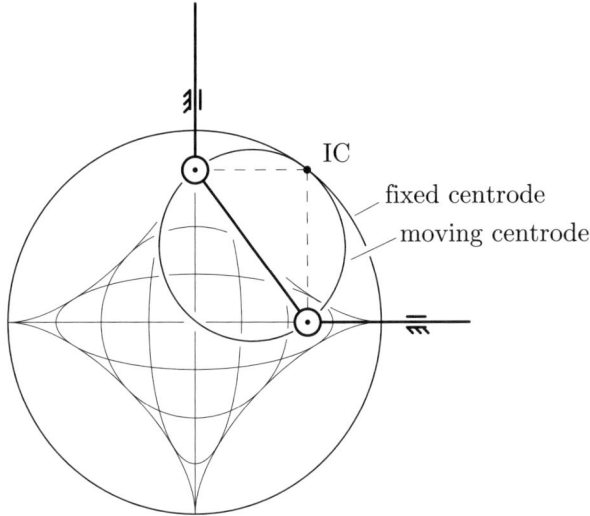

Figure 2.8
Construction of fixed and moving centrodes.

In summary, the method is:

1. Reduce the constraints to point-velocity constraints.
2. Construct perpendiculars to the allowed velocities at each point.
3. If all perpendiculars have a common intersection, that is the instantaneous center. Parallel lines are assumed to intersect at infinity. If there are three or more perpendiculars, there may be no intersection, meaning no motion is feasible.

By repeating this analysis for several different configurations of a mechanism we can construct the fixed and moving centrodes. Figure 2.8 shows example centrodes constructed for the motion of the coupler link of the four-bar linkage shown in figure 2.7. This is a particularly interesting mechanism: the centrodes are circles, points on the coupler link centerline sweep out ellipses, and the coupler link's motion sweeps out an *astroid*.

This method does have limitations. It identifies any feasible instantaneous center, but it may also identify a center which is not feasible—a *false positive*. Figure 2.9 shows an immobile five-bar linkage which under our first-order analysis appears to have a rotation center.

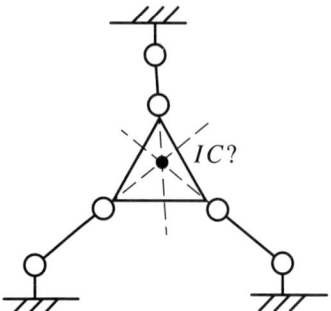

Figure 2.9
A false instantaneous center: no rigid motion is possible.

2.3 Spherical kinematics

Spherical kinematics refers to the possible motions on the surface of a sphere. Why should we care about this class of motions? Recall the definition of rotation: a displacement that leaves one point fixed. For three-space, this is equivalent to the possible motions on the surface of a sphere, the fixed point being the center of the sphere. Traditionally this is called spherical kinematics, but our interest would be better expressed by the name *spatial rotation kinematics*.

Spherical kinematics also has a surprisingly close relationship to planar kinematics. If we view the plane to be the surface of a sphere, where the radius has passed to infinity, then planar kinematics is analogous to spherical kinematics. Consequently, the results in this section are analogous to the results in the previous section.

THEOREM 2.5: A displacement of the sphere is completely determined by the motion of any two points that are not antipodal.

Proof As in the planar case, we use the two points to define a coordinate system, so that the motion of any point can be determined using its coordinates. ∎

THEOREM 2.6 EULER'S THEOREM: For every spatial rotation, there is a line of fixed points. In other words, every rotation about a point is a rotation about a line, called the *rotation axis*.

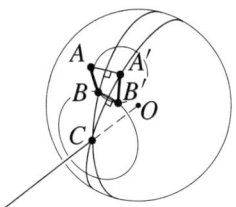

Figure 2.10
Construction for the proof of Euler's theorem.

Proof We will prove that a spherical displacement always has a fixed point on the sphere, from which the theorem follows. Let D be a given displacement of a sphere with center O. Let A be a point on the sphere, and let A' be its image under D. If $A = A'$ then we have our fixed point, so we will only consider the case where A and A' are distinct. Let $\perp AA'$ be the great circle on the sphere equidistant from A and A'. Let B be any point on $\perp AA'$, and let B' be its image under D. If $B = B'$ then once again we would have a fixed point, so we need only consider the case where B and B' are distinct. Define $\perp BB'$ to be the great circle equidistant from B and B'. The great circles $\perp AA'$ and $\perp BB'$ are distinct, since $B \in \perp AA'$ and $B \notin \perp BB'$. Hence they intersect at two antipodal points; let C be either.

Let R be the rotation mapping A to A' and mapping C to itself. If we can show that R maps B to B', then $D = R$ and the proof will be complete.

To determine $R(B)$, we find all points that satisfy the distance-preserving property of a displacement. These points have to be the right distance $|BC|$ from C, and the right distance $|BA|$ from A'. Each constraint defines a circle, one circle centered on C and one circle centered on A'. The centers of these circles A' and C, are neither identical, nor antipodes, so the intersection of the two circles has at most two points. By construction of the circle, B is the right distance from C. Then B' is also the right distance from C, by construction of C on $\perp BB'$. We also have that B is the right distance from A' by its construction on $\perp AA'$. Finally, B' is the right distance from A' since $A'B'$ is the image of AB under the given displacement D.

So, either $R(B) = B$, or $R(B) = B'$. The former can be excluded, since it would imply that R is the null motion. That cannot be the case since $R(A) = A' \neq A$. We conclude that $R(B) = B'$, hence the rotation R with fixed point C is identical to the given displacement D.

That shows that a spherical motion has a fixed point. Since the center of the sphere is also fixed, then every spatial rotation has two fixed points. The line through these two points is the desired rotation axis. It follows easily that every point on the rotation axis is fixed. ∎

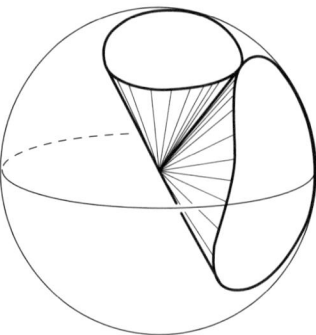

Figure 2.11
Any spherical motion is equivalent to the rolling of a moving cone on a fixed cone without slip.

The point M in the above construction is analogous to the rotation center for a planar displacement, and the proof of Euler's theorem is similar to the proof of the existence of rotation centers for planar displacements (theorem 2.4).

Since spherical kinematics is analogous to planar kinematics, we should look for an analog to the fixed and moving centrodes. Any spherical motion is equivalent to the rolling of one moving cone, without slipping, on a fixed cone with common apex.

2.4 Spatial kinematics

We now consider arbitrary displacements in space. Recall that, as in the plane, we can describe a displacement as a rotation followed by a translation. This gives us a convenient description of spatial displacements, but it is not a canonical description, since it depends on the choice of a reference point. For planar displacements, we discovered a canonical description, using the rotation center. Can the same be accomplished for spatial displacements? Unfortunately, not all spatial displacements are rotations. As an example, consider the *screw displacement* illustrated in figure 2.12: we rotate about some line in space and simultaneously translate along the same line. For a general screw motion there are no fixed points, not even at infinity. A point on the screw axis moves along it, and a point off the axis moves along a helix. Since there are no fixed points, it is not a rotation.

As the next theorem shows, however, the screw displacements exhaust the spatial displacements, and provide a nearly canonical geometric description of spatial displacements.

THEOREM 2.7 CHASLES'S THEOREM: Every spatial displacement is the composition of a rotation about some axis, and a translation along the same axis.

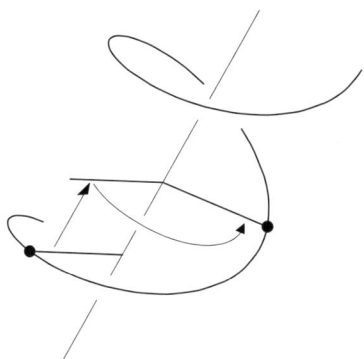

Figure 2.12
A screw displacement combines a translation and a rotation with a common axis.

Proof Let D be an arbitrary spatial displacement, and decompose it into a rotation R and a translation T, by theorem 2.2—$D = R \circ T$. Now decompose the translation T into two components $T_\perp$ and $T_\parallel$, perpendicular and parallel, respectively, to the rotation axis of R. So we now have the decomposition $D = R \circ T_\perp \circ T_\parallel$. Now $R \circ T_\perp$ is a planar motion, and is therefore equivalent to some rotation S about an axis parallel to the rotation axis of R. This yields the decomposition $D = S \circ T_\parallel$. This decomposition completes the proof, since the axis of $T_\parallel$ can be taken equal to the rotation axis of S. ∎

The proof suggests that the screw axis might sometimes be infinitely distant, and we can admit such screws if we like, but they are unnecessary. The proof would construct a screw axis at infinity only in the case of a pure translation. But a pure translation can be represented trivially by a screw motion whose rotation part is zero, and whose axis is any line parallel to the translation.

Screws provide such a powerful description that a great deal of the kinematics literature is phrased in the language of screw theory, providing an endless source of titillating puns. We will not delve very deeply into screw theory, and I will eschew the puns, but we will use screws where convenient.

DEFINITION 2.9: A **screw** is a line in space with an associated pitch, which is a ratio of linear to angular quantities.

Think of a screw as a geometrical object, which can represent motion, as it does here, but which can also represent other things, force and torque in particular. A parenthetical note: we must be intentionally vague about the pitch, about whether to divide linear by

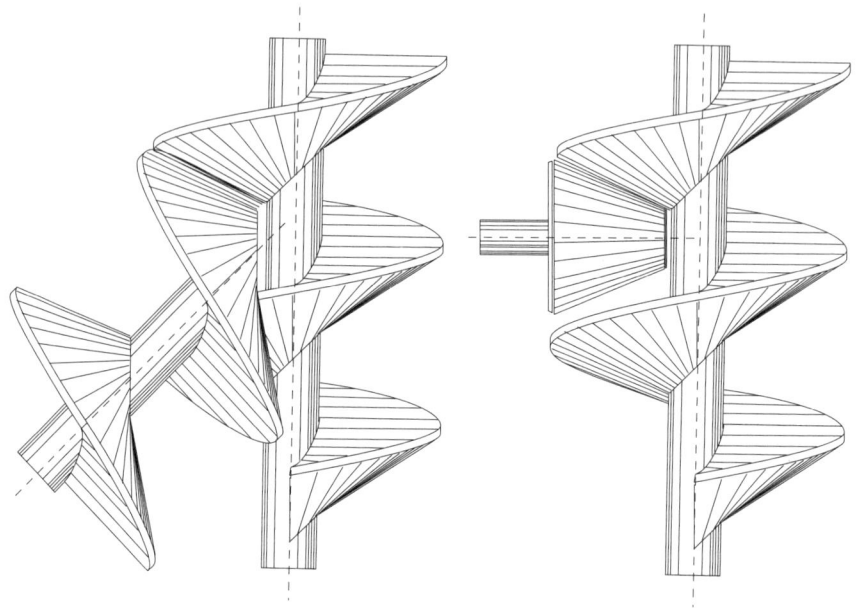

Figure 2.13
Any spatial motion is equivalent to the motion of a moving axode on a fixed axode. The instantaneous screw axis is a line of contact between the two surfaces.

angular, or the other way around. The choice depends on the application, so that we will have linear over angular quantities for motion, but angular over linear quantities for force. For further discussion of this point see section 5.1.

DEFINITION 2.10: A **twist** is a screw plus a scalar magnitude, giving a rotation about the screw axis plus a translation along the screw axis. The rotation angle is the twist magnitude, and the translation distance is the magnitude times the pitch. Thus the pitch is the ratio of translation to rotation.

Using the language of screws, Chasles's theorem is succinctly stated: every spatial displacement is a twist about some screw.

The screw is perfectly well-defined for infinitesimal motions. Hence any motion at a given time will have a screw and associated twist magnitude, describing the instantaneous velocity and angular velocity of the motion. This defines the *instantaneous screw axis*.

Just as planar motions can be described by the motion of centrodes, spatial motions are described by the motion of a *moving axode* on a *fixed axode*. Each axode is a ruled

Kinematics

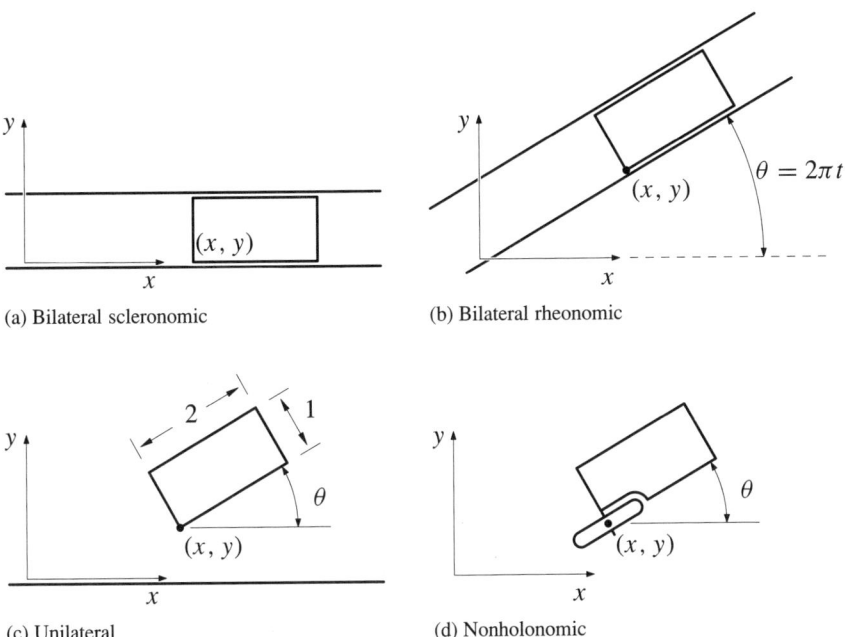

Figure 2.14
Different types of kinematic constraint.

surface. At any particular time the axodes are in contact along some common line which defines the instantaneous screw axis. The moving axode might slide along the line as it rolls, according to the pitch.

Unfortunately the motion of axodes is less easily visualized than the motion of centrodes. Of course, centrodes (figure 2.8) and cones (figure 2.11) can be viewed as special cases. Figure 2.13 shows two less trivial cases adapted from (Reuleaux, 1876).

2.5 Kinematic constraint

Manipulation usually involves contact, which can often be modeled in terms of *kinematic constraint*: a constraint on the possible motions of a body. Consider the example of a rectangular block sliding in a channel (figure 2.14(a)). Ordinarily the block would have three degrees of freedom, corresponding to free variations in the coordinates x, y, and θ. But the channel imposes a constraint, described by the equations:

$$y = 0 \tag{2.1}$$
$$\theta = 0 \tag{2.2}$$

This is the simplest kind of constraint. There are many variations, and they all have names.

For instance, suppose the rectangular channel were mounted on a turntable, rotating about the origin at a rate of one revolution per second (figure 2.14(b)). The corresponding constraint equations would be

$$x \sin(2\pi t) - y \cos(2\pi t) = 0 \tag{2.3}$$
$$\theta = 2\pi t \tag{2.4}$$

The technical name for a moving constraint is *rheonomic*. When it is necessary to make a distinction, the stationary constraint is called a *scleronomic* constraint.

Another common variation is when the constraint on motion is not symmetric, i.e. when motion is constrained in one direction but not the opposite direction. Suppose that we remove one of the channel walls in our example (figure 2.14(c)). We can describe the constraint by looking at each vertex in turn, writing an inequality that keeps the vertex in the positive half-plane:

$$y \geq 0 \tag{2.5}$$
$$y + 2\sin\theta \geq 0 \tag{2.6}$$
$$y + 2\sin\theta + \cos\theta \geq 0 \tag{2.7}$$
$$y + \cos\theta \geq 0 \tag{2.8}$$

Although we now have four constraint inequalities, note that at most two can be *active* at any time, because at most two vertices can be in contact with the constraint surface. The accepted term for a one-sided constraint is *unilateral*. When it is necessary to make a distinction, the two-sided variation is called a *bilateral* constraint.

The final variation is the most interesting. Suppose that we remove both walls of the channel, but add a wheel to the block, so it behaves like a unicycle or an ice skate (figure 2.14(d)). At any given point in time, the block can move forward and backward, it can rotate about the wheel center, but it cannot move sideways. We can describe this constraint by the equation:

$$\dot{x} \sin\theta - \dot{y} \cos\theta = 0 \tag{2.9}$$

The difference is that this equation involves the *rate* variables $\dot{x}$ and $\dot{y}$, rather than just the configuration variables x, y, and θ. Of course, we can differentiate any of the previous cases to obtain equations involving rate variables. However, the unicyclist's constraint is *not* so obtained—it is impossible to describe this constraint using just configuration

Kinematics

variables. For this reason, the constraints are often described as *nonintegrable* constraints, or as *nonholonomic* constraints.

Is a unilateral constraint nonholonomic? After searching through the dustiest and most respectable looking applied mechanics texts I can find, I am satisfied that a *holonomic* constraint should be defined as one that can be described as an equation in configuration variables and time, $F(\mathbf{q}, t) = 0$, and a *nonholonomic* constraint should be defined as one that cannot. For a nonholonomic constraint, either rate variables or inequalities are required. That means a unilateral constraint is nonholonomic. But be aware that the robotics literature often neglects this point, and in fact this book will henceforth also neglect it.

Recall that *degrees of freedom* is defined to be the number of independent variables required to determine the system configuration. Then it is evident that each independent holonomic constraint reduces the degrees of freedom of the system by one, but a nonholonomic constraint does not. This distinction is an important one, and is addressed in more detail in the next section.

The different types of kinematic constraint can be summarized as follows:

bilateral two-sided constraint, which can be expressed by equations of the form $F(\ldots) = 0$.
unilateral a one-sided constraint, requiring an inequality $F(\ldots) \geq 0$.
holonomic a constraint that can be expressed as an equation in just the configuration variables, and possibly time, but independent of the rate variables, $F(\mathbf{q}, t) = 0$.
nonholonomic a constraint that cannot be expressed in the form $F(\mathbf{q}, t) = 0$, requiring either inequalities or rate variables.
scleronomic a stationary constraint, expressible independent of time $F(\mathbf{q}, \dot{\mathbf{q}}) = 0$.
rheonomic a moving constraint, involving time $F(\mathbf{q}, \dot{\mathbf{q}}, t) = 0$.

Nonholonomic constraints

How do we know if a kinematic constraint is nonholonomic? Given a constraint equation using rate variables

$$f(\mathbf{q}, \dot{\mathbf{q}}, t) = 0$$

how do we know whether the same constraint can be written without the rate variables?

For example, consider the example of a wheeled cart in figure 2.15. We can write two constraint equations

$$\dot{x} \sin\theta - \dot{y} \cos\theta = 0 \qquad (2.10)$$

$$\dot{\theta} = 0 \qquad (2.11)$$

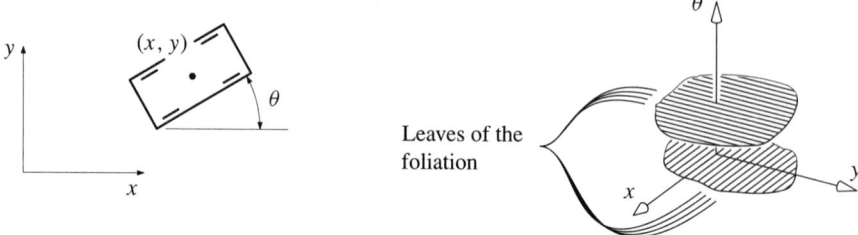

Figure 2.15
An unsteered cart, once placed, must live on a single line in configuration space. These lines comprise a *foliation* of the configuration space.

which involve rate variables and thus *appear* to describe nonholonomic constraints. But we could also have written

$$(x - x_0) \sin \theta - (y - y_0) \cos \theta = 0 \tag{2.12}$$
$$\theta - \theta_0 = 0 \tag{2.13}$$

which reveals the constraints to be holonomic. We can see that the constraints are holonomic by looking at configuration space. Equations 2.12 and 2.13 restrict the cart's configuration to a line in configuration space. There are several different possible lines, corresponding to different initial placements of the cart (x_0, y_0, θ_0). For a particular choice of θ_0, the corresponding x-y plane of configuration space is covered by parallel lines at angle θ_0 (figure 2.15). By varying the choice of θ_0, we tile the entire configuration space with lines. Each line represents the reduced 1-freedom configuration space. This decomposition of configuration space into subspaces is called a *foliation* of configuration space, and each line is a *leaf* of the foliation.

What about the unicycle? Is it possible that a similar foliation can be produced? For the unicycle, it is easy to construct a motion to any configuration (x, y, θ): turn until the wheel points at (x, y); roll forward to (x, y); turn to angle θ. Thus we know that the configuration space is truly three-dimensional, and that the constraint cannot be written as an equality constraint on the configuration variables alone.

Thus there is a fundamental difference between the cart example of figure 2.15 and the unicycle example of figure 2.14(d). The cart's constraints equations can be written without using rate variables, i.e. they are *integrable*. The unicycle's equations, on the other hand, are nonintegrable, i.e. truly nonholonomic.

To tell whether a system is holonomic, geometric reasoning sometimes suffices as in the examples above. There is also an analytical method, using *Lie brackets*. First we must introduce some terminology. Let **C** be the configuration space, and write the configuration

Kinematics

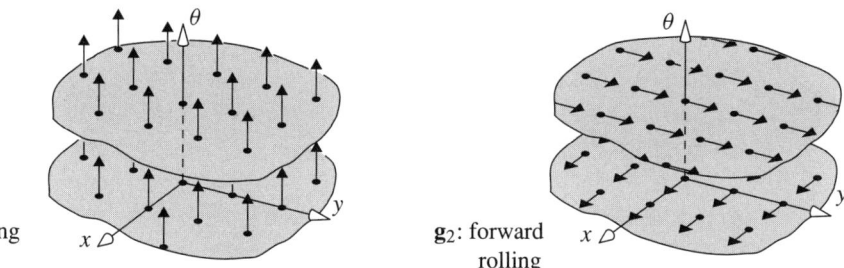

g_1: turning g_2: forward rolling

Figure 2.16
A unicycle is capable of two independent motions described by two vector fields.

as $\mathbf{q} \in \mathbf{C}$. $\mathbf{T_q C}$ is the *tangent space* at $\mathbf{q}$, the space of all velocity vectors, which can be visualized as a copy of $\mathbf{R}^n$ with its origin placed at $\mathbf{q}$. A velocity vector is written $\dot{\mathbf{q}} \in \mathbf{T_q C}$.

DEFINITION 2.11: A set of k **Pfaffian constraints** are of the form

$$\mathbf{w}_i(\mathbf{q})\dot{\mathbf{q}} = 0, i = 1 \ldots k$$

where the $\mathbf{w}_i$ are linearly independent row vectors, and $\dot{\mathbf{q}}$ is a column vector.

DEFINITION 2.12: A **vector field** is a smooth map

$$f(\mathbf{q}) : \mathbf{C} \mapsto \mathbf{T_q C}$$

from configurations $\mathbf{q}$ to velocity vectors $\dot{\mathbf{q}}$.

Let's construct some example vector fields for the unicycle of figure 2.14(d). For this example we have $\mathbf{q} = (x, y, \theta)^T$ and $\dot{\mathbf{q}} = (\dot{x}, \dot{y}, \dot{\theta})^T$. For any given $\mathbf{q}$, there are two motions which we will use as basis motions. First, we can always rotate about the contact point:

$$\begin{pmatrix} \dot{x} \\ \dot{y} \\ \dot{\theta} \end{pmatrix} = \begin{pmatrix} 0 \\ 0 \\ 1 \end{pmatrix}$$

Second, we can roll forward in direction θ:

$$\begin{pmatrix} \dot{x} \\ \dot{y} \\ \dot{\theta} \end{pmatrix} = \begin{pmatrix} \cos\theta \\ \sin\theta \\ 0 \end{pmatrix}$$

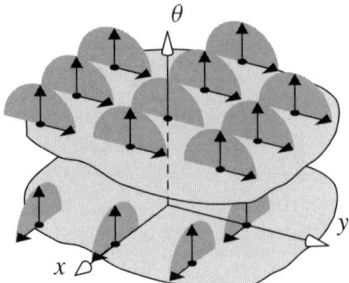

Figure 2.17
The linear span of the vector fields gives a *distribution*, the set of all feasible motions.

Thus we define two vector fields $\mathbf{g}_1$ and $\mathbf{g}_2$ (figure 2.16):

$$\mathbf{g}_1(\mathbf{q}) = \begin{pmatrix} 0 \\ 0 \\ 1 \end{pmatrix} \tag{2.14}$$

$$\mathbf{g}_2(\mathbf{q}) = \begin{pmatrix} \cos\theta \\ \sin\theta \\ 0 \end{pmatrix} \tag{2.15}$$

For this example the single Pfaffian constraint can be written

$$\mathbf{w}_1 = (\sin\theta, -\cos\theta, 0)$$

and it is easily confirmed that the products $\mathbf{w}_1 \mathbf{g}_1$ and $\mathbf{w}_1 \mathbf{g}_2$ are zero, showing that the corresponding motions are consistent with the constraint equation.

DEFINITION 2.13: A **distribution** is a smooth map assigning a linear subspace of $\mathbf{T_q C}$ to each configuration $\mathbf{q}$ of $\mathbf{C}$.

Suppose the configuration space C has dimension n. Given k Pfaffian constraints, at any $\mathbf{q}$ there will be an $(n-k)$-dimensional linear subspace of allowable velocities. Thus, returning to our example of the unicycle, we can define the distribution

$$\Delta = \text{span}(\mathbf{g}_1, \mathbf{g}_2)$$

Thus at each configuration $\mathbf{q}$, we have constructed two velocities corresponding to the two feasible motion directions, and have used those as a vector basis to construct a plane of allowable velocities. Figure 2.17 shows a few such planes, represented by small circular patches.

Kinematics

DEFINITION 2.14: A distribution is **regular** if its dimension is constant over the configuration space.

DEFINITION 2.15: Let **f**, **g** be two vector fields on **C**. Define the **Lie bracket** [**f**, **g**] to be the vector field

$$\frac{\partial \mathbf{g}}{\partial \mathbf{q}} \mathbf{f} - \frac{\partial \mathbf{f}}{\partial \mathbf{q}} \mathbf{g}$$

The two partial derivatives in the definition of Lie brackets are derivatives of the vector field with respect to a change in the configuration, and are represented by n by n matrices.

DEFINITION 2.16: A distribution is **involutive** if it is closed under Lie bracket operations.

DEFINITION 2.17: The **involutive closure** of a distribution Δ is the closure $\overline{\Delta}$ of the distribution under Lie bracketing.

THEOREM 2.8 FROBENIUS'S THEOREM: A regular distribution is integrable if and only if it is involutive.

We forego a detailed proof of Frobenius's theorem, but the method of the proof is enlightening. To prove that an integrable distribution is involutive, we consider the following maneuver: given two vector fields **f** and **g** in the distribution,

1. Follow **f** for some time ϵ;
2. Follow **g** for ϵ;
3. Follow $-\mathbf{f}$ for ϵ;
4. Follow $-\mathbf{g}$ for ϵ.

Now if you take a Taylor series expansion of this motion, the first order terms cancel, but the second order terms result in the cross-partial that we have already defined as the Lie bracket. For that reason this maneuver is often called a *Lie bracket motion*. If the distribution is integrable, then this Lie bracket motion must also be contained in the distribution, implying that the distribution is involutive. (See (Murray et al., 1994) for details.) The proof of the converse, that involutive distributions are integrable, is by induction on the dimension of the spaces. See (Boothby, 1975) for details.

Frobenius's theorem gives us a straightforward test to determine whether a system is nonholonomic. Returning to the example of the unicycle, we have a distribution

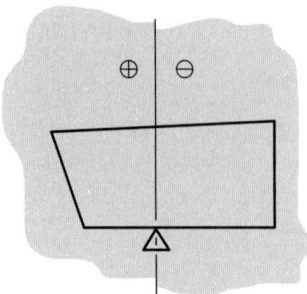

Figure 2.18
Reuleaux's method for analysis of unilateral constraints. For an IC to the right of the contact normal, only negative rotations are possible. For an IC to the left of the contact normal, only positive rotations are possible.

$\Delta = \text{span}(\mathbf{g}_1, \mathbf{g}_2)$ where the vector fields $\mathbf{g}_1$ and $\mathbf{g}_2$ are given by equations 2.14 and 2.15. The partials are

$$\frac{\partial \mathbf{g}_1}{\partial \mathbf{q}} = \begin{pmatrix} 0 & 0 & 0 \\ 0 & 0 & 0 \\ 0 & 0 & 0 \end{pmatrix} \qquad (2.16)$$

$$\frac{\partial \mathbf{g}_2}{\partial \mathbf{q}} = \begin{pmatrix} 0 & 0 & -\sin\theta \\ 0 & 0 & \cos\theta \\ 0 & 0 & 0 \end{pmatrix} \qquad (2.17)$$

For the new vector field defined by the Lie bracket we obtain

$$\mathbf{g}_3 = [\mathbf{g}_1, \mathbf{g}_2] \qquad (2.18)$$

$$= \frac{\partial \mathbf{g}_2}{\partial \mathbf{q}} \mathbf{g}_1 - \frac{\partial \mathbf{g}_1}{\partial \mathbf{q}} \mathbf{g}_2 \qquad (2.19)$$

$$= \begin{pmatrix} -\sin\theta \\ \cos\theta \\ 0 \end{pmatrix} \qquad (2.20)$$

The Lie bracket motion $\mathbf{g}_3$ is a parallel-parking maneuver, yielding a sideways motion of the unicycle, which would violate the constraint and is certainly not in the distribution $\Delta = \text{span}(\mathbf{g}_1, \mathbf{g}_2)$. Thus we see that Δ is not involutive, and so by Frobenius's theorem the unicycle is a nonholonomic system.

Kinematics

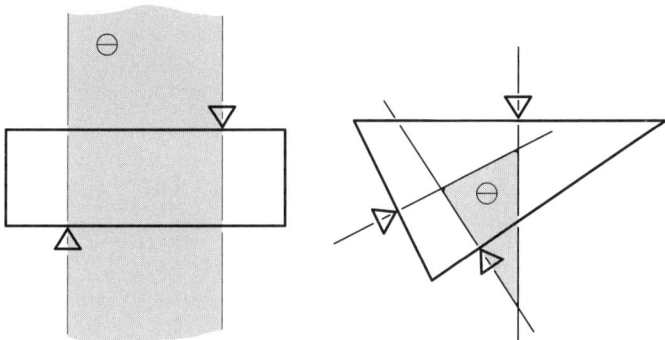

Figure 2.19
Reuleaux's method applied to several constraints. Only consistently-labeled IC's are retained.

Analysis of planar constraints by velocity centers

Bilateral planar constraints can be analyzed by identifying the feasible velocity centers, as when we constructed centrodes for four-bar linkages. There is a generalization of the method to unilateral constraints, originally described by Reuleaux. Reuleaux noticed that by simply attaching a sign to each rotation center, it is possible to analyze unilateral constraints. We will begin with the simplest possible case: a single unilateral contact (figure 2.18). For a velocity pole to the right of the contact normal, only negative rotations are possible. Similarly, only positive rotations are possible about centers to the left of the normal. For a rotation pole on the contact normal, rotations of either direction are possible. So we can describe the constraint by labeling regions in the plane of possible rotation centers either $+$, $-$, or $\pm$.

How do we describe a system involving several constraints? We simply label the regions for each individual constraint, and keep only the consistently labeled regions. In the case of two anti-parallel constraints (figure 2.19(a)), the only possible rotations are about centers falling between the two contact normals, with a negative sign. Similarly, a system of three constraints (figure 2.19(b)) typically yields a triangle of possible rotation centers, which sometimes degenerates to a point.

To summarize, Reuleaux's method is:

1. Construct a contact normal for each contact;
2. Label regions in the plane either $+$, $-$, or $\pm$ with respect to each normal;
3. Every consistently labeled region gives a set of possible rotation poles.

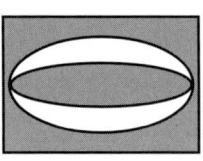 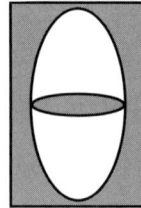

Figure 2.20
Reuleaux's method is a first-order analysis, and sometimes gives *false positives*.

It is important to be aware of the limits of this method. As with our analysis of bilateral constraints, this is a first order analysis which sometimes gives *false positives*. It may give motion centers when an object is immobile. For example, the two problems in figure 2.20 yield an identical analysis under Reuleaux's method, although one of them actually cannot move. Despite its limitations, the method is simple enough to be a useful tool, especially when aided by common sense.

2.6 Kinematic mechanisms

This section provides a brief introduction to kinematic mechanisms. A *kinematic mechanism* consists of several rigid bodies called *links* which are connected at *joints*. A joint imposes one or more constraints on the relative motion of the two links connected. The *lower pairs* (figure 2.21) comprise the special class of joints that can be constructed by two surfaces with positive contact area. Thus a cylindrical shaft in a matching cylindrical hole is a lower pair (the *cylindrical pair*), but a cylinder on a plane has just a line contact and is a *higher pair*.

The study of kinematic mechanisms addresses the problem of determining the possible motions of a given linkage, as well as the problem of designing a linkage to produce a desired motion. We have already seen some examples (figures 2.7 and 2.8). Some other interesting linkages are shown in exercises 2.5 through 2.7.

One of the main issues is to determine the *mobility* of a mechanism, defined to be the number of freedoms of the linkage where one link is assumed to be fixed. We will also employ the concept of *connectivity*, defined to be the number of freedoms of one particular link relative to another. Thus, in figure 2.22 the mobility is two ($M = 2$), and the connectivity of link two relative to link one is one ($C_{21} = 1$).

There is a simple formula for determining the mobility of a linkage. Let n be the number of links, and let g be the number of joints. For the ith joint, let u_i be the number of constraints and let f_i be the number of freedoms, and note that $u_i + f_i = 6$. If we consider

Kinematics

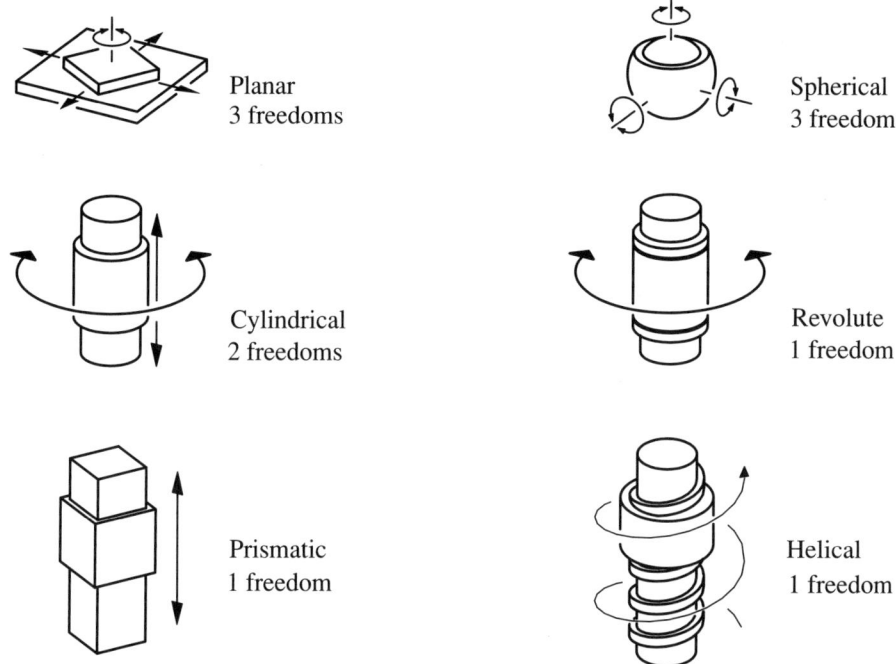

Figure 2.21
The lower pairs

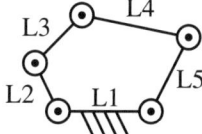

Figure 2.22
Mobility and connectivity

one link to be fixed and assume the constraints to be independent, then the mobility M is readily seen to be

$$M = 6(n-1) - \sum u_i \tag{2.21}$$
$$= 6(n-1) - \sum (6 - f_i) \tag{2.22}$$
$$= 6(n - g - 1) + \sum f_i \tag{2.23}$$

which is known as *Grübler's formula* for spatial linkages. Similarly we can obtain:

$$M = 3(n-1) - \sum u_i \qquad (2.24)$$
$$= 3(n-g-1) + \sum f_i \qquad (2.25)$$

for planar linkages. Applying the spatial variant of Grübler's formula to a planar mechanism would generally give the wrong answer, due to the dependencies among the joint constraints of a planar linkage in three space.

Another variant of Grübler's formula is applied to mechanisms with loops. First note that for a single-loop chain there are equal numbers of links and joints, so

$$M = \sum f_i + 6(-1)$$

Now if we create a two-loop mechanism by adding an open chain of k links and $k+1$ joints, we will have

$$M = \sum f_i + 6(-2)$$

Each time we add an open chain, we increase the excess of joints over links by 1. Thus for a chain comprising l loops, we have

$$M = \sum f_i - 6l$$

for a spatial linkage, and

$$M = \sum f_i - 3l$$

for a planar linkage. As an example, for a four-bar linkage we have four joints, each joint has a single freedom, and there is one loop, so that $M = 1$.

A final caution: because Grübler's formula makes such strong assumptions about the independence of constraints, it should be applied with copious amounts of common sense.

2.7 Bibliographic notes

(Reuleaux, 1876) and (Hilbert and Cohn-Vossen, 1952) are essential reading. Much of the material on kinematic linkages, and some historical background, come from (Hartenberg and Denavit, 1964). (Bottema and Roth, 1979) and (McCarthy, 1990) provide more detailed treatments of theoretical kinematics. In particular, see their analytic proofs of Euler's theorem and planar rotation poles. (Lin and Burdick, 2000) address metrics on the special Euclidean groups. The material on analysis of nonholonomic constraint is adapted from (Murray et al., 1994). Also see (Brockett, 1990) for related topics. The introduction

Kinematics

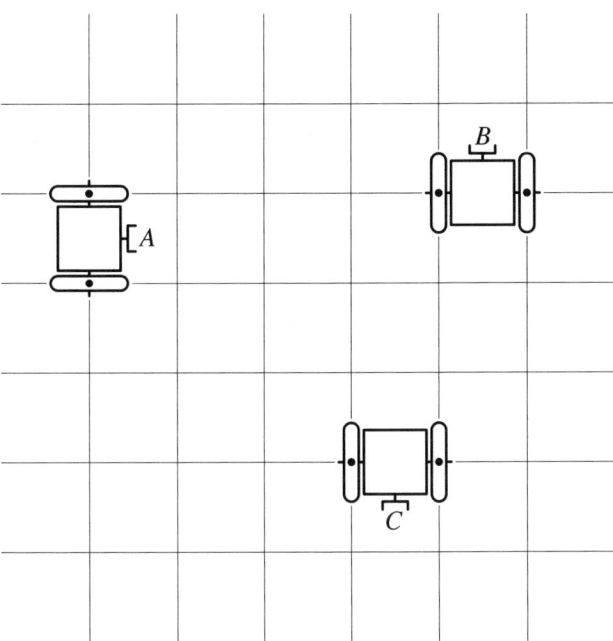

Figure 2.23
Construction for exercise 2.3.

to kinematic constraint is adapted from (Paul, 1979). For a more detailed development of screw theory and its application to mechanisms, see (Ball, 1900) and (Hunt, 1978). A higher-order analysis of kinematic constraint is given by (Rimon and Burdick, 1995).

Exercises

Exercise 2.1: Generally people say that a line in $\mathbf{E}^3$ has four freedoms, because it takes four numbers to specify a line. However, a careful interpretation of definitions 2.2 and 2.4 suggests that a line in $\mathbf{E}^3$ has five freedoms. Show that a line in $\mathbf{E}^3$ can be specified with four numbers, and explain why, according to our definitions, it has five freedoms.

Exercise 2.2: Figure 2.4 gives an example showing that spatial rotations do not commute in general. Give an example illustrating that planar displacements do not commute in general.

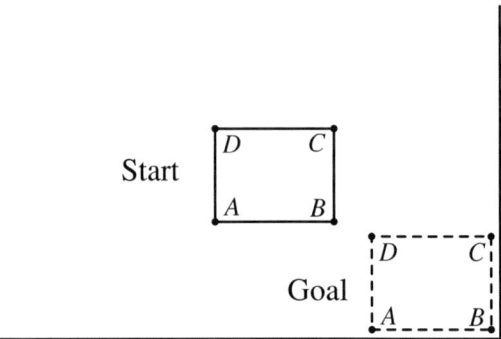

Figure 2.24
Construction for exercise 2.4.

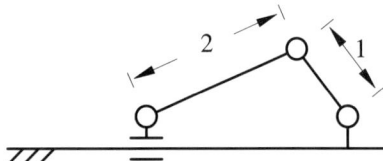

Figure 2.25
A slider-crank linkage for exercise 2.5.

Exercise 2.3: Figure 2.23 shows a planar mobile robot that cycles repeatedly through a sequence of three positions A, B, and C. The robot plans a sequence of three rotations that would cycle it through the three desired positions. Construct the fixed and moving centrodes for this cyclical motion.

The robot finds itself unable to execute the planned motion. What is the difficulty?

Exercise 2.4: Your new refrigerator has been delivered, and has to be moved from the center of your kitchen into the corner. (See figure 2.24.) You can "walk" it by shifting the weight towards one leg, then pushing so the refrigerator rotates about that leg. Find a short sequence of rotations to walk the refrigerator into the corner, and construct the centrodes. Don't go through any walls. (Hint: assembly problems are often easier to solve backwards, so find a path from the goal to the start.)

Exercise 2.5: Figure 2.25 shows a simple planar mechanism called a *slider-crank*. The slider oscillates as the crank rotates through 360 degrees. Carefully draw the mechanism in a number of configurations, and for each construct the instantaneous velocity center

Kinematics

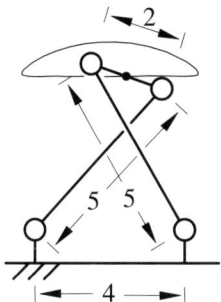

Figure 2.26
The Chebyshev linkage for exercise 2.6.

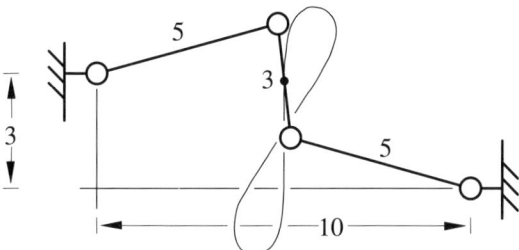

Figure 2.27
Watt's linkage for exercise 2.7.

of the coupler link. Sketch the centrodes. You will need to construct at least 12 different configurations, by sampling the crank angle at 30 degree intervals. To construct the moving centrode, you will probably need to use acetate or tracing paper.

Exercise 2.6: Figure 2.26 shows a four-bar linkage called the *Chebyshev linkage*. Two of the links rock back and forth as the coupler link makes a more complicated motion. The center of the coupler link approximates a straight line over part of its path. Construct the fixed and moving centrodes using the procedure of exercise 2.5.

Exercise 2.7: Figure 2.27 shows a simple planar mechanism called *Watt's linkage*. As in the previous exercise, the center of the coupler link approximates a straight line over part of its path. Construct the fixed and moving centrodes using the procedure of exercise 2.5.

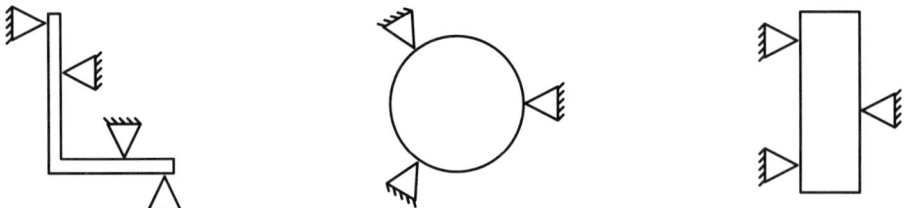

Figure 2.28
Three unilateral constraint problems for exercise 2.8.

Exercise 2.8: Apply Reuleaux's method to the problems drawn in figure 2.28, identifying all possible motions of the constrained rigid body.

Exercise 2.9: Suppose we have placed three fingers on a triangle as in figure 2.19(b). Before picking up the triangle, we want the triangle immobilized relative to the hand. Find a placement for a fourth finger that immobilizes the triangle. Apply Reuleaux's technique to prove your placement works, assuming that the fingers are perfectly stiff.

Exercise 2.10: Apply Frobenius's theorem to show that a rigid body in the plane with two independent Pfaffian constraints must be holonomic.

3 Kinematic Representation

This chapter presents representations of spatial rotations and spatial displacements. Representations are necessary for computational purposes, but more importantly, they enrich our intuitions and give us insights into the properties of spatial kinematics.

3.1 Representation of spatial rotations

There are a great number of different schemes for representing rotations, but only a very few of them are fundamentally distinct. This section presents the basic ideas, and how these ideas are reflected in different representations.

There are two big problems in representing rotations, both related to inherent, incontrovertible properties of rotations:

- Rotations do not commute. (See figure 2.4.)
- The topology of spatial rotations does not permit a smooth embedding in Euclidean three space.

The first problem, the non-commutativity of rotations, is certainly well known, and is discussed in elementary physics texts. Nonetheless, it is important to fix this fact carefully in one's mind, because some representations seem to contradict this fact (exercise 3.8). The second problem, the lack of a smooth embedding in Euclidean three-space, means that there is no smooth representation using three numbers. The problem is similar to that of assigning coordinates to locations on the surface of the Earth. Our use of longitude and latitude becomes very awkward at the poles, where a single step can produce a radical change in longitude. We don't look for a superior system because there is no such system— there just is not any way to smoothly wrap a sphere with a plane. Similarly, there is no way to smoothly wrap the space of rotations $SO(3)$ with Euclidean three space.

So in designing a representation of rotations, we have a choice: use only three numbers and suffer the resulting singularities; or use four (or more) numbers, and suffer the redundancy. The choice depends on factors that vary with the application. For computers, the redundancy is not really a problem, so most algorithms use representations with extra numbers. People, on the other hand, sometimes have a preference for the minimum set of numbers. Because of these and other differences, there is no single superior representation, and often several different representations must be maintained with procedures for translating among them.

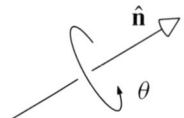

Figure 3.1
Representing spatial rotations by axis and angle.

Axis-angle

Euler's theorem (theorem 2.6) states that every spatial rotation leaves some line fixed: the rotation axis. Let us fix an origin somewhere on the rotation axis, and let $\hat{\mathbf{n}}$ be a unit vector directed along the rotation axis. Let θ be the magnitude of the rotation, with the positive sense taken in the right-hand direction about $\hat{\mathbf{n}}$. Then the ordered pair $(\hat{\mathbf{n}}, \theta)$ can indicate a rotation, which we will denote rot$(\hat{\mathbf{n}}, \theta)$. Note that the representation is two-to-one for most rotations—rot$(-\hat{\mathbf{n}}, -\theta)$ gives the same rotation as rot$(\hat{\mathbf{n}}, \theta)$. An additional source of redundancy is that rot$(\hat{\mathbf{n}}, \theta + 2k\pi)$ is identical to rot$(\hat{\mathbf{n}}, \theta)$, for any integer k. Both of these can be ameliorated to some extent by restricting θ to some suitable range, such as $[0, \pi]$. A more troublesome difficulty is that when $\theta = 0$, the rotation axis is indeterminate, giving an infinity-to-one mapping.

There are three things to do with a representation of rotation. First, we can use it simply to communicate or remember the rotation. Second, we can use it to rotate things—to compute the representation of a rotated point for example. Third, given two rotations, we might want to represent their composition. However, the axis-angle representation is a poor one for computing compositions.

To rotate a point, we will use *Rodrigues's formula* (figure 3.2). Suppose that some point to be rotated is represented by the vector $\mathbf{x}$. First, we decompose $\mathbf{x}$ into components parallel and perpendicular, respectively, to the rotation axis $\hat{\mathbf{n}}$: $\mathbf{x} = \mathbf{x}_\parallel + \mathbf{x}_\perp$. We can rewrite $\mathbf{x}_\parallel$ as $\hat{\mathbf{n}}(\hat{\mathbf{n}} \cdot \mathbf{x})$, and $\mathbf{x}_\perp$ as $-\hat{\mathbf{n}} \times (\hat{\mathbf{n}} \times \mathbf{x})$, yielding

$$\mathbf{x} = \hat{\mathbf{n}}(\hat{\mathbf{n}} \cdot \mathbf{x}) - \hat{\mathbf{n}} \times (\hat{\mathbf{n}} \times \mathbf{x}) \tag{3.1}$$

The parallel component is unaffected by the rotation. When the perpendicular component is rotated, we obtain:

$$\mathbf{x}' = \hat{\mathbf{n}}(\hat{\mathbf{n}} \cdot \mathbf{x}) + \sin\theta \ (\hat{\mathbf{n}} \times \mathbf{x}) - \cos\theta \ \hat{\mathbf{n}} \times (\hat{\mathbf{n}} \times \mathbf{x}) \tag{3.2}$$

which is *Rodrigues's formula*. A common variation is

$$\mathbf{x}' = \mathbf{x} + (\sin\theta) \ \hat{\mathbf{n}} \times \mathbf{x} + (1 - \cos\theta) \ \hat{\mathbf{n}} \times (\hat{\mathbf{n}} \times \mathbf{x}) \tag{3.3}$$

Kinematic Representation

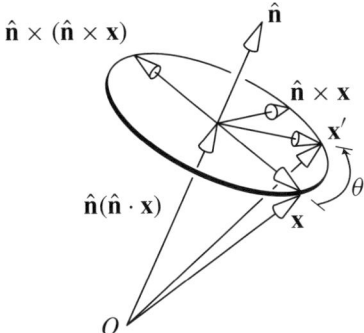

Figure 3.2
Geometric derivation of Rodrigues's formula

The uses of Rodrigues's formula go well beyond the rotation of points. For example, later sections will use Rodrigues's formula to derive transformations from one representation to another.

Rotation matrices

The *rotation matrix* is for many purposes the most useful representation of spatial rotations, because rotations of a point, and the composition of two rotations, are implemented using matrix multiplication. We begin with a derivation of the rotation matrix.

Let the origin lie on the rotation axis, and let $(\hat{\mathbf{u}}_1, \hat{\mathbf{u}}_2, \hat{\mathbf{u}}_3)$ describe a right-handed coordinate system; that is, let the $\hat{\mathbf{u}}_i$ be mutually orthogonal unit vectors with $\hat{\mathbf{u}}_1 \times \hat{\mathbf{u}}_2 = \hat{\mathbf{u}}_3$. Let $(\hat{\mathbf{u}}_1', \hat{\mathbf{u}}_2', \hat{\mathbf{u}}_3')$ be the image under the rotation. The rotation is completely determined by the motions of the $\hat{\mathbf{u}}_i$. We express the $\hat{\mathbf{u}}_i'$ vectors in $\hat{\mathbf{u}}_i$ coordinates, and collect them in a matrix:

$$\hat{\mathbf{u}}_1' = \begin{pmatrix} a_{11} \\ a_{21} \\ a_{31} \end{pmatrix} = \begin{pmatrix} \hat{\mathbf{u}}_1 \cdot \hat{\mathbf{u}}_1' \\ \hat{\mathbf{u}}_2 \cdot \hat{\mathbf{u}}_1' \\ \hat{\mathbf{u}}_3 \cdot \hat{\mathbf{u}}_1' \end{pmatrix} \tag{3.4}$$

$$\hat{\mathbf{u}}_2' = \begin{pmatrix} a_{12} \\ a_{22} \\ a_{32} \end{pmatrix} = \begin{pmatrix} \hat{\mathbf{u}}_1 \cdot \hat{\mathbf{u}}_2' \\ \hat{\mathbf{u}}_2 \cdot \hat{\mathbf{u}}_2' \\ \hat{\mathbf{u}}_3 \cdot \hat{\mathbf{u}}_2' \end{pmatrix} \tag{3.5}$$

$$\hat{\mathbf{u}}_3' = \begin{pmatrix} a_{13} \\ a_{23} \\ a_{33} \end{pmatrix} = \begin{pmatrix} \hat{\mathbf{u}}_1 \cdot \hat{\mathbf{u}}_3' \\ \hat{\mathbf{u}}_2 \cdot \hat{\mathbf{u}}_3' \\ \hat{\mathbf{u}}_3 \cdot \hat{\mathbf{u}}_3' \end{pmatrix} \tag{3.6}$$

$$A = (a_{ij}) = (\hat{\mathbf{u}}_1' | \hat{\mathbf{u}}_2' | \hat{\mathbf{u}}_3')$$

Since a rotation matrix has nine numbers, and spatial rotations have only three degrees of freedom, we have six excess numbers, and six constraints that hold among the nine numbers.

$$|\hat{\mathbf{u}}_1'| = |\hat{\mathbf{u}}_2'| = |\hat{\mathbf{u}}_3'| = 1 \qquad (3.7)$$

$$\hat{\mathbf{u}}_3' = \hat{\mathbf{u}}_1' \times \hat{\mathbf{u}}_2' \qquad (3.8)$$

which just restates that the vectors are unit vectors forming a right-handed coordinate system. Matrices satisfying these properties are called *orthonormal*, or when used to represent rotations, they are simply called *rotation matrices*.

ROTATING A POINT USING ROTATION MATRICES

If we represent a point **x** by its coordinates (x_1, x_2, x_3) in the $(\hat{\mathbf{u}}_1, \hat{\mathbf{u}}_2, \hat{\mathbf{u}}_3)$ coordinate frame, then the rotated point **x**′ is given by the same coordinates taken in the $(\hat{\mathbf{u}}_1', \hat{\mathbf{u}}_2', \hat{\mathbf{u}}_3')$ frame:

$$\begin{align}
\mathbf{x}' &= x_1 \hat{\mathbf{u}}_1' + x_2 \hat{\mathbf{u}}_2' + x_3 \hat{\mathbf{u}}_3' & (3.9) \\
&= x_1 A \hat{\mathbf{u}}_1 + x_2 A \hat{\mathbf{u}}_2 + x_3 A \hat{\mathbf{u}}_3 & (3.10) \\
&= A(x_1 \hat{\mathbf{u}}_1 + x_2 \hat{\mathbf{u}}_2 + x_3 \hat{\mathbf{u}}_3) & (3.11) \\
&= A\mathbf{x} & (3.12)
\end{align}$$

so rotating a point is implemented by ordinary matrix multiplication.

CHANGE OF COORDINATES USING ROTATION MATRICES

Rotation of a point by a rotation matrix is intimately related to the problem of change of coordinates. Suppose we have two different coordinate frames, A and B. We will use a notation common in the mechanisms and robotics literature to indicate the coordinate frame: a pre-superscript indicates the coordinate frame for a vector or a matrix. Hence:

x a point

x   a geometrical vector, directed from an origin O to the point x; or, a vector of three numbers, representing x in an unspecified frame

$^A\mathbf{x}$ a vector of three numbers, representing x in the A frame

Let $^A\mathbf{x}$ be the coordinates of some point x, taken in coordinate frame A, and let $^B\mathbf{x}$ be the coordinates of the same point, taken in coordinate frame B. Let $^B_A R$ be the rotation matrix that rotates frame B to frame A. Then we have already seen that $^B_A R$ represents the rotation of the point x:

$$^B\mathbf{x}' = {^B_A R} \, ^B\mathbf{x} \qquad (3.13)$$

Kinematic Representation

Here, we can view the pre-superscript of the rotation matrix as indicating the coordinate frame of the matrix. The operation should only be applied when the matrix and column vector are represented in the same coordinate frame, i.e. have the same pre-superscript.

However, by the same act of matrix multiplication, we can also represent the change of coordinates

$$^B\mathbf{x} = {}^B_A R \, {}^A\mathbf{x} \tag{3.14}$$

Here the pre-superscript of the vector should match the pre-subscript of the matrix. Intuitively, these are "canceled out" by the multiplication, leaving only the pre-superscript B.

COMPOSITION OF ROTATIONS USING ROTATION MATRICES

Let the rotation matrices R_1 and R_2 represent two rotations to be taken in succession. Let p be some point, p' its image after the first rotation, and p'' its image after the second rotation. Here the matrices and column vectors are taken in the same unspecified coordinate frame.

$$\mathbf{p}' = R_1 \mathbf{p} \tag{3.15}$$
$$\mathbf{p}'' = R_2(\mathbf{p}') \tag{3.16}$$
$$= R_2(R_1 \mathbf{p}) \tag{3.17}$$
$$= (R_2 R_1)\mathbf{p} \tag{3.18}$$

by associativity of matrix multiplication. Hence the composition of two rotations is represented by the product of the two rotation matrices.

OTHER PROPERTIES OF ROTATION MATRICES

Rotation matrices are often the representation of choice, because of some properties that simplify calculations:

- The null rotation is represented by the identity matrix:

$$\text{rot}(\hat{\mathbf{n}}, 0) \mapsto I \tag{3.19}$$

- The inverse of a rotation is given by matrix transpose:

$$\text{rot}(\hat{\mathbf{n}}, -\theta) \mapsto R^{-1} = R^T \tag{3.20}$$

- To change coordinates of a rotation matrix:

$$^A R = {}^A_B R \, {}^B R \, {}^B_A R \tag{3.21}$$

where $^A R$, $^B R$, are rotation matrices, and $^A_B R$, $^B_A R$ are coordinate transform matrices.

CONVERTING AXIS-ANGLE TO MATRIX REPRESENTATIONS

Suppose we have a rotation represented by the axis-angle method, and we need a rotation matrix. Rodrigues's formula provides a straightforward approach to this problem, which can be very tedious by other methods. Rodrigues's formula is:

$$\mathbf{x}' = \mathbf{x} + (\sin\theta)\,\hat{\mathbf{n}} \times \mathbf{x} + (1 - \cos\theta)\,\hat{\mathbf{n}} \times (\hat{\mathbf{n}} \times \mathbf{x})$$

A useful trick in these situations is to rewrite vector cross-product as a matrix operation. Define N:

$$N = \begin{pmatrix} 0 & -n_3 & n_2 \\ n_3 & 0 & -n_1 \\ -n_2 & n_1 & 0 \end{pmatrix} \tag{3.22}$$

so that

$$N\mathbf{x} = \hat{\mathbf{n}} \times \mathbf{x} \tag{3.23}$$

Substituting into Rodrigues's formula:

$$\mathbf{x}' = \mathbf{x} + (\sin\theta)N\mathbf{x} + (1 - \cos\theta)N^2\mathbf{x} \tag{3.24}$$

Factoring $\mathbf{x}$ from the right hand side yields an expression for the rotation matrix

$$R = I + (\sin\theta)N + (1 - \cos\theta)N^2 \tag{3.25}$$

Expanding the rotation matrix yields

$$\begin{pmatrix} n_1^2 + (1-n_1^2)c\theta & n_1 n_2(1-c\theta) - n_3 s\theta & n_1 n_3(1-c\theta) + n_2 s\theta \\ n_1 n_2(1-c\theta) + n_3 s\theta & n_2^2 + (1-n_2^2)c\theta & n_2 n_3(1-c\theta) - n_1 s\theta \\ n_1 n_3(1-c\theta) - n_2 s\theta & n_2 n_3(1-c\theta) + n_1 s\theta & n_3^2 + (1-n_3^2)c\theta \end{pmatrix} \tag{3.26}$$

where $c\theta = \cos\theta$ and $s\theta = \sin\theta$.

CONVERTING MATRIX TO AXIS-ANGLE REPRESENTATIONS

Given some rotation matrix R, we need an axis-angle representation $\text{rot}(\hat{\mathbf{n}}, \theta)$. First, we should note that when $\theta = 0$ the axis is undetermined, so we should not expect to compute $\hat{\mathbf{n}}$ when R is the identity matrix. Further, we must be content with a method that is ill-conditioned as R approaches the identity matrix. It is in the nature of the problem that arbitrarily small changes in R produce significant changes in $\hat{\mathbf{n}}$, when R is close to the identity matrix.

On the other hand, there is no such difficulty when $\theta = 180$. Some widely known methods fail in this case, so beware. One simple method is to convert the matrix to a quaternion and then convert the quaternion to axis-angle form. Well-behaved methods for each of these conversions are given in section 3.1.

DIFFERENTIAL ROTATIONS

Consider Rodrigues's formula for a differential rotation $rot(\hat{\mathbf{n}}, d\theta)$.

$$\mathbf{x}' = (I + \sin d\theta N + (1 - \cos d\theta)N^2)\mathbf{x} \qquad (3.27)$$
$$= (I + d\theta N)\mathbf{x} \qquad (3.28)$$

so

$$d\mathbf{x} = N\mathbf{x}\, d\theta \qquad (3.29)$$
$$= \hat{\mathbf{n}} \times \mathbf{x}\, d\theta \qquad (3.30)$$

giving a simple justification of the use of cross-product for differential rotations. If we define the angular velocity vector $\boldsymbol{\omega}$:

$$\boldsymbol{\omega} = \hat{\mathbf{n}} \frac{d\theta}{dt} \qquad (3.31)$$

then we obtain

$$d\mathbf{x} = \boldsymbol{\omega} \times \mathbf{x}\, dt \qquad (3.32)$$

It follows easily that differential rotations are vectors: you can scale them and add them up.

SUMMARY OF MATRIX REPRESENTATION

The rotation matrix is a convenient representation for many reasons: the mapping between rotation matrices and spatial rotations is one-to-one, with no singularities. Rotating a vector is simple. Composition of rotations is simple, and finding an inverse rotation is simple. The familiarity of matrix algebra is another feature. The primary disadvantage is that there are so many numbers, which often make rotation matrices inscrutable. Numerical errors may build up until it is necessary to normalize a rotation matrix, using singular value decomposition or other techniques.

Euler angles

Spatial rotations can be represented by three numbers, which give the angles of three rotations taken in succession about axes chosen from some coordinate frame. Several different conventions are in use, which vary in the choice of axes, and also in whether the succeeding rotations are about the *transformed* axes, or about the original *fixed* axes.

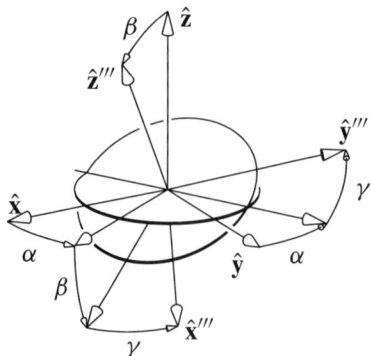

Figure 3.3
Representing spatial rotations by Euler angles.

This section adopts the ZYZ convention: the Euler angles (α, β, γ) are interpreted as a rotation of α about the z axis, then a rotation of β about the (transformed) y' axis, and then a rotation of γ about the (twice transformed) z'' (figure 3.3):

$$(\alpha, \beta, \gamma) \mapsto \text{rot}(\gamma, \hat{\mathbf{z}}'') \, \text{rot}(\beta, \hat{\mathbf{y}}') \, \text{rot}(\alpha, \hat{\mathbf{z}}) \qquad (3.33)$$

Any spatial rotation can be expressed using ZYZ Euler angles. Suppose we are given some coordinate frame $(\hat{\mathbf{x}}, \hat{\mathbf{y}}, \hat{\mathbf{z}})$ and its image under the desired rotation $(\hat{\mathbf{x}}''', \hat{\mathbf{y}}''', \hat{\mathbf{z}}''')$. We can express the Euler angles as follows:

Rotate α about $\hat{\mathbf{z}}$ until $\hat{\mathbf{y}}' \perp \hat{\mathbf{z}}'''$;
Rotate β about $\hat{\mathbf{y}}'$ until $\hat{\mathbf{z}}'' \parallel \hat{\mathbf{z}}'''$;
Rotate γ about $\hat{\mathbf{z}}''$ until $\hat{\mathbf{y}}'' = \hat{\mathbf{y}}'''$.

Note that the procedure does not completely determine the parameters α, β, and γ. In the general case where $\sin \beta \neq 0$, there are two choices, since $(\alpha + \pi, -\beta, \gamma + \pi)$ gives the same result as (α, β, γ). Two special cases occur when $\sin \beta = 0$. If $\hat{\mathbf{z}} = \hat{\mathbf{z}}'''$ then $\beta = 0$, and α may be chosen freely—only the sum of α and γ matters. If $\hat{\mathbf{z}} = -\hat{\mathbf{z}}'''$, then $\beta = \pi$, and again α may be chosen freely—only the difference of α and γ matters. So the mapping from Euler angles to spatial rotations is generally two to one, except for two cases where the mapping is a continuum to one.

Euler angles are not a convenient representation for rotating points, nor for constructing composite rotations. It is generally wiser to transform to rotation matrices or some other representation better suited to computational uses.

Transforming Euler Angles to Rotation Matrix

ZYZ Euler angles map to a spatial rotation by the formula

$$\mathrm{rot}(\gamma, \hat{\mathbf{z}}'') \, \mathrm{rot}(\beta, \hat{\mathbf{y}}') \, \mathrm{rot}(\alpha, \hat{\mathbf{z}}) \tag{3.34}$$

Fortunately, we can reverse the order, and take the rotations about fixed axes (see exercise 3.3).

$$\mathrm{rot}(\alpha, \hat{\mathbf{z}}) \, \mathrm{rot}(\beta, \hat{\mathbf{y}}) \, \mathrm{rot}(\gamma, \hat{\mathbf{z}}) \tag{3.35}$$

We can obtain the equivalent rotation matrix by substituting rotation matrices for each factor above, then expanding the product. We adopt the notation $c\alpha = \cos\alpha$, $s\alpha = \sin\alpha$, etc.

$$\begin{pmatrix} c\alpha & -s\alpha & 0 \\ s\alpha & c\alpha & 0 \\ 0 & 0 & 1 \end{pmatrix} \begin{pmatrix} c\beta & 0 & s\beta \\ 0 & 1 & 0 \\ -s\beta & 0 & c\beta \end{pmatrix} \begin{pmatrix} c\gamma & -s\gamma & 0 \\ s\gamma & c\gamma & 0 \\ 0 & 0 & 1 \end{pmatrix}$$

$$= \begin{pmatrix} c\alpha\, c\beta\, c\gamma - s\alpha\, s\gamma & -c\alpha\, c\beta\, s\gamma - s\alpha\, c\gamma & c\alpha\, s\beta \\ s\alpha\, c\beta\, c\gamma + c\alpha\, s\gamma & -s\alpha\, c\beta\, s\gamma + c\alpha\, c\gamma & s\alpha\, s\beta \\ -s\beta\, c\gamma & s\beta\, s\gamma & c\beta \end{pmatrix} \tag{3.36}$$

The transformation in the other direction—from a rotation matrix to a set of Euler angles—is not quite so straightforward. Suppose we are given some matrix

$$(r_{ij}) = \begin{pmatrix} r_{11} & r_{12} & r_{13} \\ r_{21} & r_{22} & r_{23} \\ r_{31} & r_{32} & r_{33} \end{pmatrix} \tag{3.37}$$

We set (r_{ij}) equal to the matrix 3.36 and solve for α, β, and γ, in terms of the r_{ij}. It would be simple to find α by $\tan^{-1}(r_{23}, r_{13})$, and to find γ by $\tan^{-1}(r_{32}, -r_{31})$. (We assume $\tan^{-1}$ is a two-argument arctangent, which maps coordinates of a point (y, x) to the appropriate angle.) However, this method does not address the special cases that occur when $\sin\beta = 0$. It may seem that we should treat those cases separately, but instead we will use a more elegant method that handles all cases uniformly. Note, however, this method requires an arctangent routine that does not generate an error at $\tan^{-1}(0, 0)$.

The main idea is to work with the sum and difference of α and γ. Let σ be the sum and δ the difference.

$$\sigma = \alpha + \gamma \tag{3.38}$$
$$\delta = \alpha - \gamma \tag{3.39}$$

Then

$$\alpha = (\sigma + \delta)/2 \qquad (3.40)$$
$$\gamma = (\sigma - \delta)/2 \qquad (3.41)$$

Now, turning our attention to the matrix 3.36 we observe

$$r_{22} + r_{11} = \cos\sigma(1 + \cos\beta) \qquad (3.42)$$
$$r_{22} - r_{11} = \cos\delta(1 - \cos\beta) \qquad (3.43)$$
$$r_{21} + r_{12} = \sin\delta(1 - \cos\beta) \qquad (3.44)$$
$$r_{21} - r_{12} = \sin\sigma(1 + \cos\beta) \qquad (3.45)$$

To solve for σ and δ we write

$$\sigma = \tan^{-1}(r_{21} - r_{12}, r_{22} + r_{11}) \qquad (3.46)$$
$$\delta = \tan^{-1}(r_{21} + r_{12}, r_{22} - r_{11}) \qquad (3.47)$$

There still appears to be a problem at $\sin\beta = 0$, but in fact this approach resolves it neatly. At $\beta = 0$, we should get a solution for σ, but δ is undetermined. At $\beta = \pi$, we should get a solution for δ, but σ is undetermined. That is precisely the behavior of this solution. The undetermined values will default to whatever $\tan^{-1}(0, 0)$ returns, often 0. Away from the singularities, both σ and δ are uniquely determined. Given σ and δ we use equations 3.40 and 3.41 to obtain α and γ. To obtain β, we use the solution for α to compute $\cos\alpha$ and $\sin\alpha$, then use

$$\beta = \tan^{-1}(r_{13}\cos\alpha + r_{23}\sin\alpha, r_{33}) \qquad (3.48)$$

SUMMARY OF EULER ANGLE REPRESENTATION

The main convenience of Euler angles is that they use only three numbers. There is no redundancy. They provide a good way of visualizing a spatial rotation. Also, they are used in the dynamic analysis of spinning bodies. But for most purposes other methods are preferable.

Quaternions

A quaternion is a 4-tuple of reals, with operations of addition and multiplication defined according to rules that will be described presently. The quaternion was introduced by Hamilton, as a generalization of complex numbers. Just as complex numbers enable us to multiply and divide two-dimensional vectors, quaternions enable us to multiply and divide four-dimensional vectors. And, just as complex multiplication implements rotation of the

Kinematic Representation

plane, quaternion multiplication implements rotation of a four-dimensional space. With a simple trick we can also use it to rotate three-dimensional space.

Besides being an elegant construction, quaternions are useful in the representation of rotations. The elements of a quaternion are also known as the *Euler parameters* of a finite rotation, not to be confused with the Euler angles. (If that is not confusing enough, Cheng and Gupta (1989) state that Euler was actually the first to derive Rodrigues's formula, while Rodrigues was first to derive the Euler parameters. The final twist is that, as Altmann (1989) tells it, Gauss had already invented quaternions but never bothered to publish them.)

DEFINITION 3.1: A **real quaternion** is a 4-tuple (q_0, q_1, q_2, q_3), sometimes written in terms of four basis elements:

$$q = q_0 1 + q_1 i + q_2 j + q_3 k \tag{3.49}$$

where the q_i are all real numbers. The following operations are defined:

1. The **sum** of two quaternions is like vector sum:

$$p + q = (p_0 + q_0)1 + (p_1 + q_1)i + (p_2 + q_2)j + (p_3 + q_3)k \tag{3.50}$$

2. The **product** of a scalar and a quaternion is given by:

$$wq = (wq_0)1 + (wq_1)i + (wq_2)j + (wq_3)k, \quad w \in \mathbf{R} \tag{3.51}$$

3. Quaternion **product** is defined by stipulating that multiplication distribute over addition, and specifying the products of the basis elements:

$$i^2 = j^2 = k^2 = -1 \tag{3.52}$$
$$ij = k \tag{3.53}$$
$$jk = i \tag{3.54}$$
$$ki = j \tag{3.55}$$

4. Quaternion **conjugate** is analogous to complex conjugate:

$$q^* = q_0 1 - q_1 i - q_2 j - q_3 k \tag{3.56}$$

5. Quaternion **length** is defined to be

$$|q| = \sqrt{qq^*} = \sqrt{q_0^2 + q_1^2 + q_2^2 + q_3^2} \tag{3.57}$$

It is easily shown that addition and multiplication have the right properties—addition is associative and commutative, multiplication is associative but not commutative.

A quaternion can also be interpreted as the sum of a *scalar part* q_0 and a *vector part* **q**:

$$q = q_0 + \mathbf{q} \tag{3.58}$$

where

$$\mathbf{q} = q_1 \mathbf{i} + q_2 \mathbf{j} + q_3 \mathbf{k} \tag{3.59}$$

It is easily shown that quaternion product can be written:

$$pq = p_0 q_0 - \mathbf{p} \cdot \mathbf{q} + p_0 \mathbf{q} + q_0 \mathbf{p} + \mathbf{p} \times \mathbf{q} \tag{3.60}$$

It is legitimate to view scalars, vectors, and complex numbers as specializations of quaternions—addition and multiplication give the familiar operations. If p and q have zero scalar parts, they are pure vectors, and their product yields both the dot product and the cross product:

$$pq = p_0 q_0 - \mathbf{p} \cdot \mathbf{q} + p_0 \mathbf{q} + q_0 \mathbf{p} + \mathbf{p} \times \mathbf{q} \tag{3.61}$$

$$= -\mathbf{p} \cdot \mathbf{q} + \mathbf{p} \times \mathbf{q} \tag{3.62}$$

Note that every quaternion, other than the additive identity 0, has an inverse

$$q^{-1} = q^* / |q|^2 \tag{3.63}$$

so that quaternions form a *linear algebra* and a *field*, the only extension of the complex numbers that is both a linear algebra and a field.

Now, we consider the use of quaternions to represent spatial rotations. Given some rotation $\text{rot}(\theta, \hat{\mathbf{n}})$, define the corresponding unit quaternion to be

$$q = \cos\frac{\theta}{2} + \sin\frac{\theta}{2} \hat{\mathbf{n}} \tag{3.64}$$

Now let x be a pure vector, i.e. a quaternion with zero scalar part,

$$x = 0 + \mathbf{x} \tag{3.65}$$

and let the vector components be the Cartesian coordinates of some point. To rotate a point, we form the product qxq^*, which we can show by expanding the product, and simplifying:

$$qxq^* = (\cos\frac{\theta}{2} + \sin\frac{\theta}{2} \hat{\mathbf{n}}) \mathbf{x} (\cos\frac{\theta}{2} - \sin\frac{\theta}{2} \hat{\mathbf{n}}) \tag{3.66}$$

$$= \cos^2\frac{\theta}{2} \mathbf{x} + 2\cos\frac{\theta}{2}\sin\frac{\theta}{2} \hat{\mathbf{n}} \times \mathbf{x} + \sin^2\frac{\theta}{2} \hat{\mathbf{n}} \mathbf{x} \hat{\mathbf{n}}^* \tag{3.67}$$

Kinematic Representation

We can now substitute the half-angle formulas

$$\cos\theta = \cos^2\frac{\theta}{2} - \sin^2\frac{\theta}{2} \qquad (3.68)$$

$$\sin\theta = 2\cos\frac{\theta}{2}\sin\frac{\theta}{2} \qquad (3.69)$$

to obtain

$$qxq^* = 1 + \cos\frac{\theta}{2}\mathbf{x} + \sin\theta\,\hat{\mathbf{n}}\times\mathbf{x} + 1 - \cos\frac{\theta}{2}\hat{\mathbf{n}}\mathbf{x}\hat{\mathbf{n}}^* \qquad (3.70)$$

It is easily shown that for $\hat{\mathbf{n}}$ a unit vector and $\mathbf{x}$ a vector we have

$$\hat{\mathbf{n}}\mathbf{x}\hat{\mathbf{n}}^* = \mathbf{x} + 2\hat{\mathbf{n}}\times(\hat{\mathbf{n}}\times\mathbf{x}) \qquad (3.71)$$

Substituting above and simplifying we obtain Rodrigues's formula for the rotated point, showing that the quaternion product rotates a point.

$$qxq^* = \mathbf{x} + \sin\theta\,\hat{\mathbf{n}}\times\mathbf{x} + (1-\cos\theta)\hat{\mathbf{n}}\times(\hat{\mathbf{n}}\times\mathbf{x}) \qquad (3.72)$$

A GEOMETRIC VIEW OF ROTATION BY QUATERNIONS

Despite the analytical argument above, the unit quaternion representation may still seem puzzling. It would seem natural to use expressions closer to the complex number analogy. In particular, why not use unit quaternions of the form

$$p = \cos\theta + \hat{\mathbf{n}}\sin\theta \qquad (3.73)$$

and why not use the expression

$$\mathbf{x}' = p\mathbf{x} \qquad (3.74)$$

to rotate vectors by a single quaternion multiplication? Some insight can be gained by a geometric look at quaternion multiplication. Let p be a unit quaternion with no scalar part, and consider the map $L_p(q)$, left multiplication by p.

$$L_p(q) = pq \qquad (3.75)$$

Then $L_p(q)$ is a rotation of the four-dimensional space of quaternions. This is apparent if we use the definition of quaternion product to express L_p as a matrix product:

$$L_p(q) = \begin{pmatrix} p_0 & -p_1 & -p_2 & -p_3 \\ p_1 & p_0 & -p_3 & p_2 \\ p_2 & p_3 & p_0 & -p_1 \\ p_3 & -p_2 & p_1 & p_0 \end{pmatrix} \begin{pmatrix} q_0 \\ q_1 \\ q_2 \\ q_3 \end{pmatrix} \qquad (3.76)$$

and then note that the matrix is orthonormal.

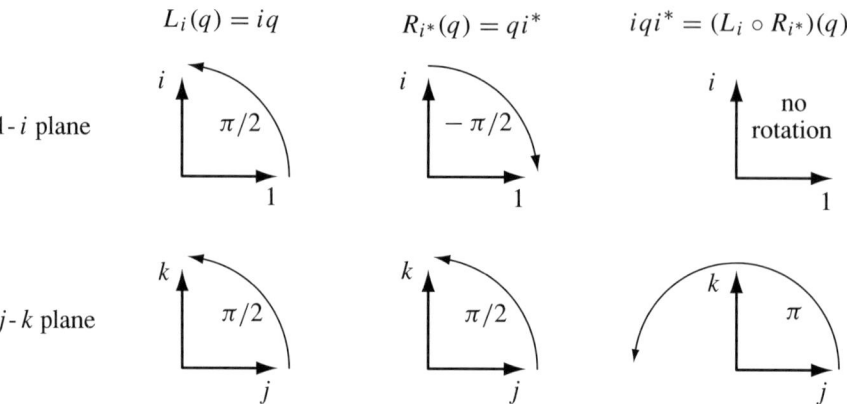

Figure 3.4
Left multiplication and right multiplication by unit quaternions, applied individually, mix up the vector and scalar subspaces. But applied together using conjugates, they give a rotation of the vector subspace.

Although left multiplication is a rotation of the four-dimensional space, it is *not* a rotation of the three-dimensional subspace of pure vectors. Left multiplication mixes up the scalar subspace and vector subspace. In fact, left multiplication by a unit pure vector p can be described as two rotations. The first rotation is a $\pi/2$ rotation of the 1-p plane, which mixes up the scalar subspace and the vector subspace. The second rotation is a $\pi/2$ rotation of the plane perpendicular to 1 and p, which involves only the vector subspace. Let's simplify things by taking p to be the basis element i.

$$L_i(q) = iq \qquad (3.77)$$

Then from the definition of quaternion product we immediately obtain:

$$L_i(q) \begin{cases} 1 \mapsto i \\ i \mapsto -1 \\ j \mapsto k \\ k \mapsto -j \end{cases} \qquad (3.78)$$

corresponding to a $\pi/2$ rotation of the 1-i plane and a $\pi/2$ rotation of the j-k plane (see figure 3.4).

Similarly, we can define $R_{p*}(q)$ to be the mapping obtained by right multiplication by a unit pure vector p^*. As before, it is a rotation of the four-dimensional quaternion space, but it rotates the 1-p plane in the opposite direction, i.e. by $-\pi/2$, while it rotates the perpendicular plane in the same direction, i.e. by $\pi/2$. Let us use the element i^* to illustrate:

$$R_{i*}(q) \begin{cases} 1 \mapsto -i \\ i \mapsto 1 \\ j \mapsto k \\ k \mapsto -j \end{cases} \tag{3.79}$$

Now it is evident that by composing $L_i(q)$ with $R_{i*}(q)$ we cause the rotations of the 1-i plane to cancel, and the rotations of the j-k plane to add up, yielding a rotation of the vector space by π about i:

$$iqi^* = (L_i \circ R_{i*})(q) \begin{cases} 1 \mapsto 1 \\ i \mapsto i \\ j \mapsto -j \\ k \mapsto -k \end{cases} \tag{3.80}$$

In general, nxn^*, for n a pure unit vector, is a rotation of the vector subspace about n by π, corresponding to the $\hat{\mathbf{n}} \times (\hat{\mathbf{n}} \times \mathbf{x})$ term in Rodrigues's construction (equation 3.71). We can construct geometric interpretations for the other elements of Rodrigues's formula in a similar fashion, ultimately obtaining a geometric interpretion of our earlier proof that the quaternion product qxq^* implements general spatial rotations.

COMPOSITION OF ROTATIONS REPRESENTED BY UNIT QUATERNIONS

The result of rotation q (represented as a unit quaternion) on some point $\mathbf{x}$ (represented as a vector) is given by

$$q\mathbf{x}q^* \tag{3.81}$$

The result after a second rotation p is given by

$$p(q\mathbf{x}q^*)p^* = (pq)\mathbf{x}(pq)^* \tag{3.82}$$

which shows that composition is accomplished using ordinary quaternion product.

TRANSFORMING BETWEEN QUATERNION AND AXIS-ANGLE REPRESENTATIONS

By convention, the quaternion representation is

$$q = \cos\frac{\theta}{2} + \sin\frac{\theta}{2}\hat{\mathbf{n}} \tag{3.83}$$

which gives the transformation from axis-angle to quaternion representation. To obtain the axis and angle from the quaternion:

$$\theta = 2\tan^{-1}(|\mathbf{q}|, q_0) \tag{3.84}$$

$$\hat{\mathbf{n}} = \mathbf{q}/|\mathbf{q}| \tag{3.85}$$

We note that the equation for the rotation axis $\hat{\mathbf{n}}$ is ill-conditioned near $\theta = 0$, but this is a result of the fundamental indeterminacy of the rotation axis for a null rotation.

TRANSFORMING BETWEEN QUATERNION AND MATRIX REPRESENTATIONS

We can turn a unit quaternion into the equivalent rotation matrix by expanding the product qxq^*:

$$qxq^* = (q_0^2 - \mathbf{q}\cdot\mathbf{q})\mathbf{x} + 2q_0\mathbf{q}\times\mathbf{x} + 2\mathbf{q}\mathbf{q}\cdot\mathbf{x} \tag{3.86}$$

$$= \left((q_0^2 - q_1^2 - q_2^2 - q_3^2)I + 2q_0\begin{pmatrix} 0 & -q_3 & q_2 \\ q_3 & 0 & -q_1 \\ -q_2 & q_1 & 0 \end{pmatrix}\right.$$

$$\left. + 2\begin{pmatrix} q_1^2 & q_1q_2 & q_1q_3 \\ q_1q_2 & q_2^2 & q_2q_3 \\ q_1q_3 & q_2q_3 & q_3^2 \end{pmatrix}\right)\mathbf{x} \tag{3.87}$$

Expanding and simplifying yields the rotation matrix:

$$\begin{pmatrix} q_0^2 + q_1^2 - q_2^2 - q_3^2 & 2(q_1q_2 - q_0q_3) & 2(q_1q_3 + q_0q_2) \\ 2(q_1q_2 + q_0q_3) & q_0^2 - q_1^2 + q_2^2 - q_3^2 & 2(q_2q_3 - q_0q_1) \\ 2(q_1q_3 - q_0q_2) & 2(q_2q_3 + q_0q_1) & q_0^2 - q_1^2 - q_2^2 + q_3^2 \end{pmatrix} \tag{3.88}$$

Now we consider how to transform a rotation matrix (r_{ij}) into an equivalent unit quaternion. First, if we take various linear combinations of the diagonal elements of the matrix in equation 3.88, we obtain:

$$q_0^2 = \frac{1}{4}(1 + r_{11} + r_{22} + r_{33}) \tag{3.89}$$

$$q_1^2 = \frac{1}{4}(1 + r_{11} - r_{22} - r_{33}) \tag{3.90}$$

$$q_2^2 = \frac{1}{4}(1 - r_{11} + r_{22} - r_{33}) \tag{3.91}$$

$$q_3^2 = \frac{1}{4}(1 - r_{11} - r_{22} + r_{33}) \tag{3.92}$$

At this point we could take square roots, but that leaves the problem of choosing the

Kinematic Representation

signs correctly. Instead, we return to the matrix for some more equations. Taking sums and differences of each pair r_{ij}, r_{ji} yields

$$q_0 q_1 = \frac{1}{4}(r_{32} - r_{23}) \tag{3.93}$$

$$q_0 q_2 = \frac{1}{4}(r_{13} - r_{31}) \tag{3.94}$$

$$q_0 q_3 = \frac{1}{4}(r_{21} - r_{12}) \tag{3.95}$$

$$q_1 q_2 = \frac{1}{4}(r_{12} + r_{21}) \tag{3.96}$$

$$q_1 q_3 = \frac{1}{4}(r_{13} + r_{31}) \tag{3.97}$$

$$q_2 q_3 = \frac{1}{4}(r_{23} + r_{32}) \tag{3.98}$$

To obtain the quaternion, use the first four equations (3.89–3.92) to find the largest q_i^2. Either sign will do for the square root. Now, whichever q_i was obtained, there are three of the remaining six equations involving that q_i, which will yield the other three components of the quaternion.

PROPERTIES OF THE UNIT QUATERNION REPRESENTATION

Quaternions are in one fundamental respect the *right* way to represent rotations. Consider some rotation $\mathrm{rot}(\hat{\mathbf{n}}, \theta)$, represented by a unit quaternion

$$\mathrm{rot}(\hat{\mathbf{n}}, \theta) \mapsto q = \cos(\theta/2) + \sin(\theta/2)\hat{\mathbf{n}} \tag{3.99}$$

Then the shortest path on the sphere joining q with the null rotation 1 has an arc-length of $\theta/2$. This implies that the spherical metric on unit quaternions corresponds to measuring spatial rotations by the rotated angle, which is exactly the right metric for spatial rotations. Of course there are two possible angles for any rotation, θ and $2\pi - \theta$, corresponding to the two quaternions q and $-q$. Henceforth we will assume that the smaller angle is used, which corresponds to choosing whichever of q or $-q$ is closest to 1.

The Euclidean metric in E^4 will also serve as a metric for spatial rotations, although it does not give the angle. We can use the quaternion length $|p - q|$ to measure the difference between two quaternions, provided that we choose antipodes yielding the smallest value.

Since the spherical metric does the right thing with quaternions, the topology must be right. Unit quaternions are 4-tuples, restricted to unit length. The unit quaternions give the surface of a sphere in four-dimensional Euclidean space. Because q and $-q$ represent the same rotation, we identify antipodes on the sphere with one another, which yields projective 3-space, $\mathbf{P}^3$. Thus spatial rotations have the topology $\mathbf{P}^3$. Unit quaternions give a smooth representation of spatial rotations with the minimum of numbers.

Another implication is that unit quaternions are ideally suited to the problem of generating a random sequence of spatial rotations. If we generate a uniform distribution on the surface of the unit sphere in E^4, we also obtain a uniform distribution of rotations. (How does one generate a uniform distribution on a sphere? See exercise 3.16.)

Normalization also works very well with quaternions. The problem is that after some numerical calculations, we may obtain a quaternion that no longer lies on the sphere. To normalize a quaternion we can simply divide it by its length.

Finally, in some applications quaternions offer superior computational efficiency (see exercise 3.10).

3.2 Representation of spatial displacements

This section explores methods for representing spatial displacements. The simplest method would be to decompose the displacement into a translation and a rotation (theorem 2.2), represent the translation by a vector, and represent the rotation by any of the methods described in the previous section. In particular, using a rotation matrix plus a vector leads us to the use of homogeneous coordinates. But there are other methods which offer some advantages, depending on the immediate problem. So following a section on homogeneous coordinates there is a section on the use of screws and screw coordinates.

Homogeneous coordinates

Recall theorem 2.2: a displacement can be decomposed into a rotation followed by a translation. We construct an origin and a coordinate frame, and represent points as coordinate vectors. Then we can represent the rotation by a rotation matrix, and represent the translation by vector addition:

$$\mathbf{x}' = R\mathbf{x} + \mathbf{d} \tag{3.100}$$

where R is the rotation matrix and d is the translation vector. This is a fairly simple equation, but it can be even simpler, using *homogeneous coordinates*. The homogeneous coordinate representation of a point is obtained by appending a fourth coordinate, which is always 1:

$$\mathbf{x} = \begin{pmatrix} x_1 \\ x_2 \\ x_3 \\ 1 \end{pmatrix} \tag{3.101}$$

Kinematic Representation

Now we construct the *homogeneous coordinate transform matrix T*:

$$T = \left(\begin{array}{c|c} R & \mathbf{d} \\ \hline 0\ 0\ 0 & 1 \end{array} \right) \tag{3.102}$$

Now the transformation of a point can be written more compactly:

$$\mathbf{x}' = T\mathbf{x} \tag{3.103}$$

Thus the homogeneous coordinate transform matrix T can represent the displacement. The first three columns of T give the rotation part, and the last column gives the translation part.

Points at infinity have a convenient representation with homogeneous coordinates. (See Appendix A for an expanded discussion.) Let w be the fourth coordinate of a point, and adopt the convention that w is a scale factor. Now our representation of points is given by:

$$\begin{pmatrix} x_1 \\ x_2 \\ x_3 \\ w \end{pmatrix} \mapsto \begin{pmatrix} x_1/w \\ x_2/w \\ x_3/w \end{pmatrix} \tag{3.104}$$

As w tends to zero, the point tends toward infinity. We adopt the convention that the homogeneous coordinate vector

$$\begin{pmatrix} x_1 \\ x_2 \\ x_3 \\ 0 \end{pmatrix} \tag{3.105}$$

represents a point at infinity. Or, equivalently, it represents a direction: the direction of all lines parallel to the vector $(x_1, x_2, x_3)^T$, which intersect at the point at infinity. Note that the points at infinity form a plane. It may appear that they form a three-dimensional space, but a point at infinity does not vary when the homogeneous coordinates are scaled.

The homogeneous coordinate representation of points at infinity gives an elegant decomposition of transform matrices. The first three columns have fourth coordinate zero. They represent points at infinity, the directions corresponding to the three coordinate axes. The fourth column has fourth coordinate one, and represents the location of the coordinate frame origin.

The main feature of homogeneous coordinates is that the transform equation is homogeneous (equation 3.103) rather than just linear (equation 3.100). ("Linear" in this context means that the plot is a straight line, and "homogeneous" means that the straight line passes through the origin. A more modern terminology might be to call them "linear coordinates" because the transform equation is linear rather than just affine.) The value of homogeneous coordinates can be better appreciated when several displacements occur in succession, which can be written

$$T_6 T_5 T_4 T_3 T_2 T_1 \tag{3.106}$$

rather than

$$R_6(R_5(R_4(\ldots) + \mathbf{d}_4) + \mathbf{d}_5) + \mathbf{d}_6 \tag{3.107}$$

Of course, we could use the simpler expression even without homogeneous coordinates, since we know that spatial displacements form a group. But it is convenient to have a simple mechanism to decompose a displacement into rotation and translation operators, and to relate the algebra to concrete numerical operations.

Naive numerical applications of homogeneous coordinates can be inefficient. It is possible to use general matrix multiplication and matrix inversion, but much more efficient procedures are obtained by taking into account the special structure of homogeneous coordinate transform matrices. Inversion of a displacement requires only a transpose of the rotation matrix, and one point transform:

$$\left(\begin{array}{c|c} R & \mathbf{d} \\ \hline 0 \ 0 \ 0 & 1 \end{array} \right)^{-1} = \left(\begin{array}{c|c} R^T & -R^T \mathbf{d} \\ \hline 0 \ 0 \ 0 & 1 \end{array} \right) \tag{3.108}$$

Composition of two displacements can also be computed more efficiently:

$$\left(\begin{array}{c|c} R_2 & \mathbf{d}_2 \\ \hline 0 \ 0 \ 0 & 1 \end{array} \right) \left(\begin{array}{c|c} R_1 & \mathbf{d}_1 \\ \hline 0 \ 0 \ 0 & 1 \end{array} \right) = \left(\begin{array}{c|c} R_2 R_1 & R_2 \mathbf{d}_1 + \mathbf{d}_2 \\ \hline 0 \ 0 \ 0 & 1 \end{array} \right) \tag{3.109}$$

Screw coordinates

The screw was introduced in section 2.4: it is a line in space, with an associated pitch. *Screw coordinates* are a method for representing screws. First, though, we must explore *Plücker coordinates*, which are used to describe lines in space.

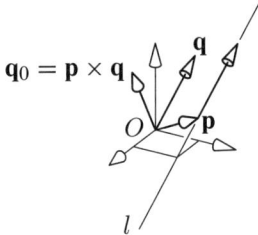

Figure 3.5
Representing a line using Plücker coordinates.

PLÜCKER COORDINATES

The equation of a line can be given in parametric form:

$$\mathbf{x}(t) = \mathbf{p} + t\mathbf{q} \tag{3.110}$$

where $\mathbf{p}$ is any point on the line, and $\mathbf{q}$ is a vector giving the direction of the line. Let $\mathbf{q}_0$ be given by:

$$\mathbf{q}_0 = \mathbf{p} \times \mathbf{q} \tag{3.111}$$

Then the ordered pair $(\mathbf{q}, \mathbf{q}_0)$ comprises the 6 *Plücker coordinates* of the line (figure 3.5). We will call $\mathbf{q}$ the *direction vector* and $\mathbf{q}_0$ the *moment vector*.

An obvious question at this point is, why use the cross product? Why not just use the point and the direction $(\mathbf{p}, \mathbf{q})$, or, even simpler, why not just use two points? We shall see soon that Plücker coordinates simplify many of the computations on lines. But there is a more basic reason: Plücker coordinates are nearly a canonical representation of lines. A line in space is determined by four numbers. That is two less than the six numbers to determine the configuration of a rigid body, because a translation along the line, or a rotation about the line, map the line back to itself. There are six Plücker coordinates, giving an excess of two. We can account for the two excess parameters as follows. First, since $\mathbf{q}_0 = \mathbf{p} \times \mathbf{q}$, any set of Plücker coordinates has to satisfy the equation

$$\mathbf{q} \cdot \mathbf{q}_0 = 0 \tag{3.112}$$

Second, we can scale the Plücker coordinates by a non-zero scalar k without changing the line:

$$(\mathbf{q}, \mathbf{q}_0) \equiv k(\mathbf{q}, \mathbf{q}_0) \tag{3.113}$$

It is sometimes convenient to normalize by scaling by $1/|\mathbf{q}|$, but we shall see cases where unnormalized Plücker coordinates are necessary.

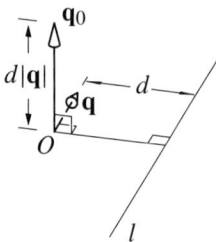

Figure 3.6
Plücker coordinates: the generic case.

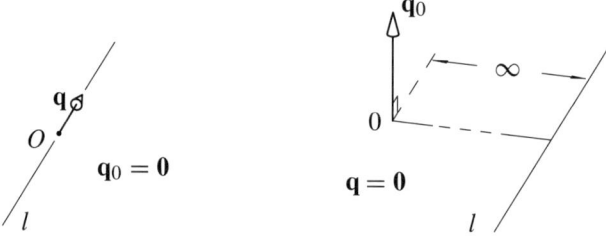

Figure 3.7
Special cases of Plücker coordinates: a line through the origin and a line at infinity.

It really is very simple to read Plücker coordinates. There are three cases:

General case. The general case (figure 3.6) is with non-zero $\mathbf{q}$ and $\mathbf{q}_0$. The direction vector $\mathbf{q}$ is parallel to the line, $\mathbf{q}_0$ is normal to the plane including the origin and the line, and $|\mathbf{q}_0|/|\mathbf{q}|$ gives the distance from the origin to the line.

Line through origin $(\mathbf{q}, \mathbf{0})$. The second case occurs when the line passes through the origin (figure 3.7(a)). This is really a straightforward instance of the general case, where $\mathbf{q}_0$ has passed to zero.

Line at infinity $(\mathbf{0}, \mathbf{q}_0)$. The third case is more interesting, and occurs when the direction vector $\mathbf{q}$ passes to zero. One way to look at it is to write the Plücker coordinates for the general case, normalized by the magnitude of the *moment* vector:

$$\left(\frac{\mathbf{q}}{|\mathbf{q}_0|}, \frac{\mathbf{q}_0}{|\mathbf{q}_0|} \right) \tag{3.114}$$

Kinematic Representation

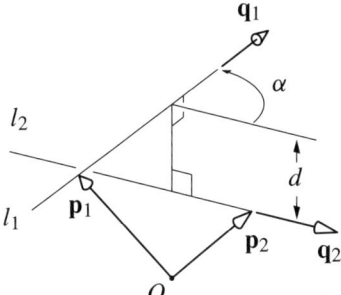

Figure 3.8
Geometrical construction of the moment between two lines.

Now consider the limiting process as the line approaches infinity. If we had normalized by the *direction* vector, then the moment vector would grow proportionately with the distance to the line. But, since we normalized by the moment vector, the direction vector will shrink in inverse proportion to the line's distance, attaining **0** in the limit.

For a line at infinity, it may seem that we have lost some vital information since we have no direction vector. But when we note that the line at infinity is the intersection of the set of all planes perpendicular to the moment vector, it is evident that the moment vector does completely specify the line.

No meaning is assigned to the Plücker coordinates $(\mathbf{0}, \mathbf{0})$.

With Plücker coordinates we can easily determine the moment between two lines, the shortest distance between two lines, and the angle between two lines. We will begin with the moment. Suppose we are given two lines l_1 and l_2 (figure 3.8). Let $\mathbf{p}_1$, $\mathbf{p}_2$ be points on l_1, l_2, respectively, and let $\mathbf{q}_1$, $\mathbf{q}_2$ be the directions of l_1, l_2, respectively. Then, in analogy with the moment of force about a point or about a line, we can define the moment of the line l_2 about the point $\mathbf{p}_1$ to be

$$(\mathbf{p}_2 - \mathbf{p}_1) \times \frac{\mathbf{q}_2}{|\mathbf{q}_2|} \tag{3.115}$$

and we can define the moment of the line l_2 about the line l_1 to be

$$\frac{\mathbf{q}_1}{|\mathbf{q}_1|} \cdot (\mathbf{p}_2 - \mathbf{p}_1) \times \frac{\mathbf{q}_2}{|\mathbf{q}_2|}, \tag{3.116}$$

which simplifies to the expression

$$\frac{\mathbf{q}_1 \cdot \mathbf{q}_{02} + \mathbf{q}_2 \cdot \mathbf{q}_{01}}{|\mathbf{q}_1||\mathbf{q}_2|}, \tag{3.117}$$

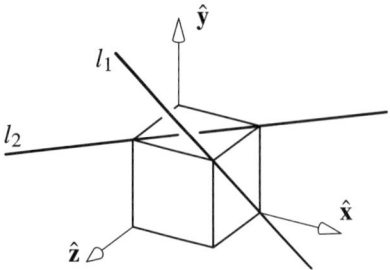

Figure 3.9
Example for Plücker coordinates.

Note that the Plücker coordinates are really giving us *directed* lines, and the expressions above are giving us the signed moment, consistent with the lines' directions. The numerator of equation 3.117 includes an operation that we will encounter again and again:

DEFINITION 3.2: We define the **virtual product**, also called **reciprocal product**, of two sets of Plücker coordinates as follows

$$(\mathbf{p}, \mathbf{p}_0) * (\mathbf{q}, \mathbf{q}_0) = \mathbf{p} \cdot \mathbf{q}_0 + \mathbf{q} \cdot \mathbf{p}_0 \tag{3.118}$$

If we have *normalized* Plücker coordinates, then virtual product gives the signed moment between two directed lines.

If d is the shortest distance between lines l_1 and l_2, and $\alpha \in [0, \pi]$ is the angle between lines l_1 and l_2, then the moment between l_1 and l_2 is given by $d \sin \alpha$. Noting that

$$|\mathbf{q}_2 \times \mathbf{q}_1| = |\mathbf{q}_1||\mathbf{q}_2| \sin \alpha \tag{3.119}$$

we can equate our two formulas for moment, yielding

$$d = \frac{|(\mathbf{q}_1, \mathbf{q}_{01}) * (\mathbf{q}_2, \mathbf{q}_{02})|}{|\mathbf{q}_2 \times \mathbf{q}_1|} \tag{3.120}$$

Note that the two lines l_1 and l_2 intersect if and only if

$$(\mathbf{q}_1, \mathbf{q}_{01}) * (\mathbf{q}_2, \mathbf{q}_{02}) = 0, \tag{3.121}$$

where parallel lines are considered to intersect at infinity. The expression for distance between two lines will not work when the lines are parallel.

EXAMPLE

Consider the diagonals of adjacent faces on a cube (figure 3.9). Line l_1 is described by

$$\mathbf{p}_1 = (1, 0, 0) \tag{3.122}$$
$$\mathbf{q}_1 = (0, 1, 1) \tag{3.123}$$

and line l_2 is described by

$$\mathbf{p}_2 = (0, 1, 1) \tag{3.124}$$
$$\mathbf{q}_2 = (-1, 0, 1) \tag{3.125}$$

To complete the Plücker coordinates of the two lines, we calculate

$$\mathbf{q}_{01} = \mathbf{p}_1 \times \mathbf{q}_1 = (0, -1, 1) \tag{3.126}$$
$$\mathbf{q}_{02} = \mathbf{p}_2 \times \mathbf{q}_2 = (1, -1, 1) \tag{3.127}$$

So the distance between the two lines is

$$d = \frac{|(\mathbf{q}_1, \mathbf{q}_{01}) * (\mathbf{q}_2, \mathbf{q}_{02})|}{|\mathbf{q}_2 \times \mathbf{q}_1|} = 1/\sqrt{3} \tag{3.128}$$

and the angle is given by

$$\alpha = \sin^{-1} \frac{|\mathbf{q}_1 \times \mathbf{q}_2|}{|\mathbf{q}_1||\mathbf{q}_2|} = \sin^{-1}(\frac{\sqrt{3}}{2}) = 60° \tag{3.129}$$

SCREW COORDINATES

Recall that a screw is a line in space, with an associated pitch. How can we assign coordinates to a screw? We could represent the screw by using the six Plücker coordinates for the line, plus a seventh number for the pitch. But recall that Plücker coordinates have an excess of numbers. We can use this excess to encode the pitch without adding a seventh number.

Consider a screw whose line is given by the Plücker coordinates $(\mathbf{q}, \mathbf{q}_0)$, and whose pitch is given by the scalar p. If the pitch is finite, we define the screw coordinates to be $(\mathbf{s}, \mathbf{s}_0)$, where

$$\mathbf{s} = \mathbf{q} \tag{3.130}$$
$$\mathbf{s}_0 = \mathbf{q}_0 + p\mathbf{q} \tag{3.131}$$

If the pitch is infinite, we make the obvious extension and define the screw coordinates to be

$$\mathbf{s} = \mathbf{0} \tag{3.132}$$
$$\mathbf{s}_0 = \mathbf{q} \tag{3.133}$$

Comparing this definition with the Plücker coordinates of a line at infinity, it appears that an infinite-pitch screw is indistinguishable from a screw with an axis at infinity, and that pitch is not meaningful for a screw axis at infinity.

For a screw with finite pitch and finite axis, since the two Plücker vectors are orthogonal ($\mathbf{q} \cdot \mathbf{q}_0 = 0$), we can recover the pitch p by taking the dot product between the two screw coordinate vectors

$$\mathbf{s} \cdot \mathbf{s}_0 = \mathbf{q} \cdot \mathbf{q}_0 + p\mathbf{q} \cdot \mathbf{q} \tag{3.134}$$
$$p = \frac{\mathbf{s} \cdot \mathbf{s}_0}{\mathbf{s} \cdot \mathbf{s}} \tag{3.135}$$

It is also straightforward to obtain the direction of the screw axis—it is simply $\mathbf{s}$. Finally, the point on the screw axis nearest the origin is given by

$$\mathbf{r} = \frac{\mathbf{s} \times \mathbf{s}_0}{\mathbf{s} \cdot \mathbf{s}} \tag{3.136}$$

SCREW COORDINATES FOR TWISTS

Recall Chasles's theorem (theorem 2.7): any spatial displacement is a twist about some screw—a translation and rotation, whose ratio is determined by the pitch. Let θ be the rotation angle, let d be translation distance, let l be the screw axis, and let p be the pitch of the screw. Then the twist is a rotation θ about the axis l, and a translation d along the axis l. The pitch p is the ratio of translation to rotation, and has dimensions of length. Finite pitch corresponds to d/θ.

To represent a displacement, we need to include the magnitude of the twist. Fortunately we still have one excess parameter to work with. For an ordinary screw (finite axis, finite pitch) we normalize the screw coordinates and scale by θ.

$$\left(\theta \frac{\mathbf{s}}{|\mathbf{s}|}, \theta \frac{\mathbf{s}_0}{|\mathbf{s}|}\right) \tag{3.137}$$

Substituting the definition of screw coordinates, we obtain

$$\left(\theta \frac{\mathbf{s}}{|\mathbf{s}|}, \theta \frac{\mathbf{s}_0}{|\mathbf{s}|}\right) = \frac{1}{|\mathbf{q}|} (\theta\mathbf{q}, \theta\mathbf{q}_0 + \theta p\mathbf{q}) \tag{3.138}$$
$$= \frac{1}{|\mathbf{q}|} (\theta\mathbf{q}, \theta\mathbf{q}_0 + d\mathbf{q}) \tag{3.139}$$

Kinematic Representation

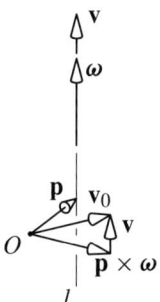

Figure 3.10
Screw coordinates for differential twists.

For a twist of infinite pitch, i.e. a translation, this definition naturally extends to

$$(\mathbf{s}, \mathbf{s}_0) = \frac{1}{|\mathbf{q}|} (\mathbf{0}, d\mathbf{q}) \tag{3.140}$$

As noted in the definition of screw coordinates, an infinite pitch screw (a translation) is indistinguishable from an infinite axis screw (a rotation about a line at infinity). Note also that the moment part of a translation axis cannot be recovered. The screw coordinates give us the direction of the axis, but not its location. This neatly reflects the nature of spatial displacements. Since a translation moves all points along parallel lines, any of these lines may be taken as the screw axis.

One other special case should be considered. A *zero pitch* screw corresponds to a twist of pure rotation, and the screw coordinates are then identical to the Plücker coordinates:

$$(\mathbf{s}, \mathbf{s}_0) = (\mathbf{q}, \mathbf{q}_0). \tag{3.141}$$

SCREW COORDINATES FOR DIFFERENTIAL TWISTS

Screw coordinates for differential twists are especially useful. As it turns out, they are also very familiar—they are identical with the use of a velocity vector and angular velocity vector. Consider the example of figure 3.10. Suppose we have a screw axis l, with an angular velocity ω about l, and a velocity of v along l. Let $\mathbf{p}$ be any point on l. The angular velocity vector $\boldsymbol{\omega}$ gives the direction of l, so the Plücker coordinates of l are

$$(\mathbf{q}, \mathbf{q}_0) = (\boldsymbol{\omega}, \mathbf{p} \times \boldsymbol{\omega}) \tag{3.142}$$

The pitch of the screw is $|\mathbf{v}|/|\boldsymbol{\omega}|$ so the screw coordinates are

$$(\mathbf{s}, \mathbf{s}_0) = (\boldsymbol{\omega}, \mathbf{p} \times \boldsymbol{\omega} + \frac{|\mathbf{v}|}{|\boldsymbol{\omega}|}\boldsymbol{\omega}) \qquad (3.143)$$

Since the velocity is parallel to the angular velocity, $(|\mathbf{v}|/|\boldsymbol{\omega}|)\boldsymbol{\omega}$ is just $\mathbf{v}$, yielding the screw coordinates

$$(\mathbf{s}, \mathbf{s}_0) = (\boldsymbol{\omega}, \mathbf{p} \times \boldsymbol{\omega} + \mathbf{v}) \qquad (3.144)$$

The second vector $\mathbf{s}_0$ is just an expression for the velocity $\mathbf{v}_0$ of a point at the origin of a globally fixed frame,

$$(\mathbf{s}, \mathbf{s}_0) = (\boldsymbol{\omega}, \mathbf{v}_0) \qquad (3.145)$$

so the use of screw coordinates for differential twists is close to the standard practice in introductory physics texts. This observation has one important corollary. *Screw coordinates for differential twists form a vector space.* We can add differential twist screw coordinates, and we can multiply them by scalars.

3.3 Kinematic constraints

This section develops the use of screw coordinates for a first-order model of kinematic constraint. Chapter 5 continues the topic, developing polyhedral convex cones and a variety of related constructs.

In the previous chapter we learned a simple graphical method (Reuleaux's method) to analyze kinematic constraint of a planar system. Screw coordinates give an analogous method in three dimensions.

Our first-order model of kinematic constraint is given by

$$\hat{\mathbf{u}} \cdot \mathbf{v}_p = 0 \qquad (3.146)$$

where $\hat{\mathbf{u}}$ is some direction in space, p is some point of the constrained body, and $\mathbf{v}_p$ is the differential motion of the point p. This is a bilateral constraint; a unilateral constraint would be represented by an inequality.

Now suppose that the differential twist of the body is given by the screw coordinates $(\boldsymbol{\omega}, \mathbf{v}_0)$, which are identical to the angular velocity of the body, and the velocity of a point at the origin of a globally fixed frame. Then the velocity of the point p is given by

$$\mathbf{v}_p = \mathbf{v}_0 + \boldsymbol{\omega} \times \mathbf{p} \qquad (3.147)$$

Kinematic Representation

So the kinematic constraint can be written

$$\hat{\mathbf{u}} \cdot (\mathbf{v}_0 + \boldsymbol{\omega} \times \mathbf{p}) = 0 \tag{3.148}$$

After some rearrangement this gives the equation

$$\hat{\mathbf{u}} \cdot \mathbf{v}_0 + (\mathbf{p} \times \hat{\mathbf{u}}) \cdot \boldsymbol{\omega} = 0 \tag{3.149}$$

which is reminiscent of the reciprocal product operation defined on Plücker coordinates. So, we define a *contact screw* describing the contact normal

$$(\mathbf{c}, \mathbf{c}_0) = (\mathbf{u}, \mathbf{p} \times \hat{\mathbf{u}}) \tag{3.150}$$

and write the kinematic constraint as

$$(\mathbf{c}, \mathbf{c}_0) * (\boldsymbol{\omega}, \mathbf{v}_0) \tag{3.151}$$

where $*$ is reciprocal product (or virtual product) extended to screw coordinates:

$$(\mathbf{s}, \mathbf{s}_0) * (\mathbf{t}, \mathbf{t}_0) = \mathbf{s} \cdot \mathbf{t}_0 + \mathbf{s}_0 \cdot \mathbf{t} \tag{3.152}$$

Note that the contact screw $(\mathbf{u}, \mathbf{p} \times \hat{\mathbf{u}})$ is a zero pitch screw, so that it is just the Plücker coordinates of the contact normal.

DEFINITION 3.3: A pair of screws is **reciprocal**, **contrary**, or **repelling**, if their reciprocal product is zero, negative, or positive, respectively.

Thus a bilateral kinematic constraint requires that the differential twist be reciprocal to the contact normal. A unilateral constraint requires that the differential twist be reciprocal or repelling to the contact normal.

The contact screw is always a zero-pitch screw. If we constrain the body twist to a pure rotation, then it also is represented by a zero-pitch screw. In this case, the reciprocal product vanishes only when the moment between the two axes vanish, that is, the screws are reciprocal only when the rotation axis intersects the contact normal. This is precisely the observation underlying Reuleaux's graphical method for analyzing planar constraint.

EXAMPLE 1

Suppose we have placed six fingers on a cube. The fingers are arranged in two groups of three, at opposite corners of a diagonal (figure 3.11). Although these are unilateral

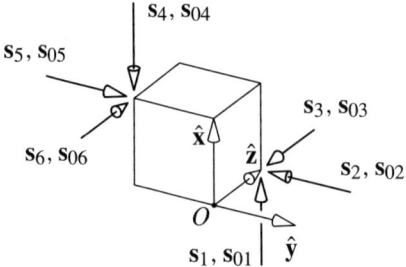

Figure 3.11
Construction for Example 1 using screw coordinates to analyze unilateral constraint.

constraints, we will consider the simpler bilateral problem. The six contact screws are

$$(\mathbf{s}_1, \mathbf{s}_{01}) = (1, 0, 0, 0, 1, 0) \tag{3.153}$$

$$(\mathbf{s}_2, \mathbf{s}_{02}) = (0, -1, 0, 1, 0, 0) \tag{3.154}$$

$$(\mathbf{s}_3, \mathbf{s}_{03}) = (0, 0, -1, 0, 0, 0) \tag{3.155}$$

$$(\mathbf{s}_4, \mathbf{s}_{04}) = (-1, 0, 0, 0, 0, -1) \tag{3.156}$$

$$(\mathbf{s}_5, \mathbf{s}_{05}) = (0, 1, 0, 0, 0, 1) \tag{3.157}$$

$$(\mathbf{s}_6, \mathbf{s}_{06}) = (0, 0, 1, -1, -1, 0) \tag{3.158}$$

Let $(\mathbf{t}, \mathbf{t}_0)$ be a differential twist consistent with the kinematic constraints. Then the reciprocal product, with each of the contact screws, must be zero:

$$\begin{aligned}
t_4   +t_2 &= 0 \\
-t_5 +t_1 &= 0 \\
-t_6 &= 0 \\
-t_4   -t_3 &= 0 \\
t_5 +t_3 &= 0 \\
t_6 -t_1 -t_2 &= 0
\end{aligned} \tag{3.159}$$

The solutions are of the form

$$(\mathbf{t}, \mathbf{t}_0) = k(1, -1, -1, 1, 1, 0) \tag{3.160}$$

for any scalar k. The pitch of this differential twist is zero, so again we have ordinary Plücker coordinates. The reader will readily verify that this is the set of differential rotations about the cube diagonal.

Kinematic Representation

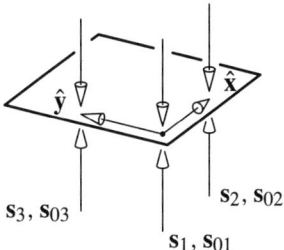

Figure 3.12
Construction for Example 2 using screw coordinates to analyze bilateral constraints.

EXAMPLE 2

How are planar motions expressed in screw coordinates? We can obtain the answer to this question by constructing an appropriate set of spatial constraints. We will choose three bilateral constraints aligned with the $\hat{\mathbf{z}}$ axis (figure 3.12). The screw coordinates for the constraints are:

$$(\mathbf{s}_1, \mathbf{s}_{01}) = (0, 0, 1, 0, 0, 0) \tag{3.161}$$

$$(\mathbf{s}_2, \mathbf{s}_{02}) = (0, 0, 1, 0, -1, 0) \tag{3.162}$$

$$(\mathbf{s}_3, \mathbf{s}_{03}) = (0, 0, 1, 1, 0, 0) \tag{3.163}$$

Let the twist be given by

$$(\mathbf{t}, \mathbf{t}_0) = (t_1, t_2, t_3, t_4, t_5, t_6) \tag{3.164}$$

The twist must be reciprocal to $(\mathbf{s}_1, \mathbf{s}_{01})$:

$$t_6 = 0 \tag{3.165}$$

to $(\mathbf{s}_2, \mathbf{s}_{02})$:

$$t_6 - t_2 = 0 \tag{3.166}$$

and to $(\mathbf{s}_3, \mathbf{s}_{03})$:

$$t_6 + t_1 = 0 \tag{3.167}$$

Thus the twist must be of the form

$$(\mathbf{t}, \mathbf{t}_0) = (0, 0, t_3, t_4, t_5, 0) \tag{3.168}$$

giving the restriction of screw coordinates to planar motions. This is a zero-pitch screw, corresponding to a pure rotation. The first three coordinates give the direction of the rotation axis—it is parallel to the $\hat{\mathbf{z}}$ axis. We can apply equation 3.136 to obtain the rotation center in the $\hat{\mathbf{x}}$-$\hat{\mathbf{y}}$ plane:

$$\frac{\mathbf{t} \times \mathbf{t}_0}{\mathbf{t} \cdot \mathbf{t}} = (-t_5 t_3, t_4 t_3, 0)/t_3^2 \tag{3.169}$$

As a special case, when $t_3 = 0$, we obtain a pure translational velocity $(t_4, t_5, 0)$.

3.4 Bibliographic notes

(Korn and Korn, 1968) provided much of the material for this chapter. (Salamin, 1979) is an excellent introduction to quaternions. (Altmann, 1989) provides some interesting historical notes on quaternions. Some of the material on Euler angles is adapted from (Crenshaw, 1994). The material on screw coordinates originated with (Roth, 1984) and (Ohwovoriole, 1980). Exercise 3.8 arose from a reading of Kane and Levinson's paper (1978). The matrix exponential provides another way to represent rotations and displacements, and yields additional insights. A good treatment is provided by (Murray et al., 1994).

Exercises

Exercise 3.1: How many degrees of freedom should be assigned to a rigid body in four dimensions? Hint: homogeneous coordinates work in any number of dimensions.

Exercise 3.2: Show correctness of the expressions given for inverse of a transform matrix and composition of two transform matrices.

Exercise 3.3: When matrices are written in transformed, rather than fixed, coordinates, the matrices are composed left-to-right, instead of right-to-left. For example, if we first rotate 30 degrees about $\hat{\mathbf{x}}$, then rotate 90 degrees about the transformed $\hat{\mathbf{y}}$, the composite transformation is given by

$$\mathbf{p}' = \text{rot}(\hat{\mathbf{y}}', 90) \, \text{rot}(\hat{\mathbf{x}}, 30)\mathbf{p} \tag{3.170}$$

However, the same rotation can be written:

$$\mathbf{p}' = \text{rot}(\hat{\mathbf{x}}, 30) \, \text{rot}(\hat{\mathbf{y}}, 90)\mathbf{p} \tag{3.171}$$

i.e. using transformed coordinates, and composing left-to-right, rather than right-to-left. Usually this is easier, but prove that it works.

Kinematic Representation

Exercise 3.4: A corollary to exercise 3.3 is that a sequence of rotations, about rotation axes taken in a fixed frame, gives the same result as if we take the same sequence rotations, reverse the order, and use the transformed frame. This might seem mind-boggling in the abstract, but is obvious when made concrete. Build a gimbal mechanism, and try out a few rotation sequences.

Exercise 3.5: Show that there are 24 different possible conventions for Euler angles.

Exercise 3.6: Repeat the analysis of section 3.1, using ZYX Euler angles: the three numbers will refer to successive rotations about $\hat{\mathbf{z}}, \hat{\mathbf{y}}', \hat{\mathbf{x}}''$.

Exercise 3.7: Euler angles are defined so that successive rotations are about orthogonal axes. Is this restriction necessary?

Exercise 3.8: An interesting paradox arises with the use of gimbal coordinates to describe spatial rotations. Suppose we have some gimbal mechanism—four links joined by revolute joints. When the mechanism is in its "home" position, the three joints lie along the $\hat{\mathbf{z}}$, $\hat{\mathbf{y}}$, and $\hat{\mathbf{z}}$ axes, respectively. Because the joints are arranged in succession, the motion of the first joint affects the location of the second and third joint axes, and the motion of the second joint affects the location of the third joint axis. The mechanism is a physical implementation of ZYZ Euler angles, mapping the joint angles to the orientation of the last link in the chain. We will refer to the three joint angles as the *gimbal coordinates* $\mathbf{g} = (g_1, g_2, g_3)$.

Now let A and B be two arbitrary orientations of the last link, and let $\mathbf{g}_A, \mathbf{g}_B$, be the corresponding gimbal coordinates. Now, for a rotation from A to B, we will assign the angles $\mathbf{g}_{\vec{AB}} = \mathbf{g}_B - \mathbf{g}_A$. Let O be the home position, represented by gimbal coordinates $(0, 0, 0)$, so that $\mathbf{g}_{\vec{OA}} = \mathbf{g}_A$. In effect we are representing spatial rotation by differences of gimbal coordinates. But consider the successive rotations $\mathbf{g}_{\vec{OA}}, \mathbf{g}_{\vec{AB}}$. You can take them in either order, since each joint just adds the numbers up. Taken in either order, the last link still ends up at B, and the composite rotation is $\mathbf{g}_{\vec{OB}}$. Apparently this method yields a *commutative* representation of spatial rotations, even though we know that rotations are not commutative. Explain this paradox. (I would claim that gimbal coordinates are not a well-defined representation of spatial rotations. So one solution would be to fill in the details in that claim.)

Exercise 3.9: The Pauli spin matrices are

$$S_1 = \begin{pmatrix} 0 & 1 \\ 1 & 0 \end{pmatrix} \tag{3.172}$$

$$S_2 = \begin{pmatrix} 0 & -i \\ i & 0 \end{pmatrix} \tag{3.173}$$

$$S_3 = \begin{pmatrix} 1 & 0 \\ 0 & -1 \end{pmatrix} \tag{3.174}$$

where $i = \sqrt{-1}$. Show that quaternions can be constructed using the two-by-two identity matrix I_2, and the three matrices $-iS_1$, $-iS_2$, and $-iS_3$, as the four elements of a vector basis.

(The Pauli spin matrices are closely related to the Cayley-Klein parameters. The construction above maps the quaternion (q_0, q_1, q_2, q_3) to the matrix

$$\begin{pmatrix} q_0 - iq_3 & -q_2 - iq_1 \\ q_2 - iq_1 & q_0 + iq_3 \end{pmatrix} \tag{3.175}$$

The Cayley-Klein parameters are the four components of the matrix.)

Exercise 3.10: Compare the computational efficiency of rotation matrices versus unit quaternions. For each method, determine the storage requirements, and the number of floating additions and multiplies required to rotate a point, and to compose two rotations.

Exercise 3.11: Construct quaternions for the null rotation, for rotations of π and $\frac{\pi}{2}$ about each coordinate axis.

Exercise 3.12: Prove that unit quaternions q and $-q$ give the same rotation.

Exercise 3.13: For the rotation described by the matrix below, produce the rotation axis and angle, the unit quaternion, and the Euler angles.

$$\begin{pmatrix} -2/3 & -2/3 & 1/3 \\ 2/3 & -1/3 & 2/3 \\ -1/3 & 2/3 & 2/3 \end{pmatrix}$$

Kinematic Representation

Exercise 3.14: For the rotation described by the matrix below, produce the rotation axis and angle, the unit quaternion, and the Euler angles.

$$\begin{pmatrix} -2/3 & 2/15 & 11/15 \\ 2/3 & -1/3 & 2/3 \\ 1/3 & 14/15 & 2/15 \end{pmatrix}$$

Exercise 3.15: Given two quaternions $p_0 + p_1 i + p_2 j + p_3 k$ and $q_0 + q_1 i + q_2 j + q_3 k$, find expressions for the components of the product $r = pq$. For example, $r_0 = p_0 q_0 - p_1 q_1 - p_2 q_2 - p_3 q_3$.

Exercise 3.16: For this question you will conduct an experiment to answer the question: What is the average angle of rotations in $\mathbf{E}^3$?

1. Write code to generate uniformly distributed unit quaternions.

2. Write code to produce the smallest rotation angle for a given quaternion, in the range from zero to π.

3. Write code to generate a lot of unit quaternions, uniformly distributed, and take the mean angle.

How does one produce a uniform distribution on the surface of a sphere in $\mathbf{E}^4$? One easy way is to generate four real numbers uniformly in the interval $[-1, 1]$. That defines a uniform distribution in a cube. If we discard every quadruple with magnitude greater than 1, we will have a uniform distribution in the interior of the sphere. Normalize to get a uniform distribution on the surface of the sphere.

Exercise 3.17: Write screw coordinates for the three contact constraints of figure 2.19(b). You can make the job easier by your choice of origin, scale, and coordinate axes.

Exercise 3.18: Consider an octahedron with vertices at $(0, 0, 1)$, $(0, 0, -1)$, $(0, 1, 0)$, $(0, -1, 0)$, $(1, 0, 0)$, and $(-1, 0, 0)$. Pick two edges that are neither intersecting nor parallel. Find the Plücker coordinates of each edge. Use reciprocal product and cross product to find the distance and angle between the two edges, as in the example on page 65.

4 Kinematic Manipulation

The goal of this chapter is to illustrate the application of kinematics to manipulation problems. First we look at *path planning*—determining a collision-free motion for a manipulator. Second we take a brief look at planning for *nonholonomic systems*. Third we look at *manipulation in the hand*—manipulating a grasped object by coordinated finger motions.

4.1 Path planning

The role of path planning is most apparent when you consider a style of manipulation we will call *pick and place* manipulation. To make things as simple as possible, suppose we have a set of blocks, all the same size, each in a known start location. We assume routines `nextblock()` which returns a data structure describing the next block to be moved; `start(block)` which returns the initial location of `block`; and `goal(block)` which returns the goal location of `block`. Pick and place manipulation is described by the code:

```
FOR block = nextblock()
    MOVETO start(block)      ; pick block
    CLOSE
    MOVETO goal(block)       ; place block
    OPEN
```

This style of manipulation makes several assumptions:

- `CLOSE` attaches the object rigidly to the effector;
- `OPEN` detaches the object;
- Unattached objects do not move;
- The robot follows the path exactly.

In practice these assumptions hold only because the engineers and programmers work furiously to enforce them. Some of the problems are:

- Grasp planning: designing the effector, and choosing effector and finger motions that will produce a stable grasp for each object.
- Stable placement: assuring that each object will be stable after it is released.
- Compliant motion: due to inevitable errors in sensing and control, placing an object involves a sequence of collisions and compliant motions, which must not damage anything.

All of these problems are simple if we are just moving blocks from one isolated location to another, without turning them over. In other tasks these problems are quite challenging.

We will mainly stick with naive assumptions, and neglect the problems of grasp planning, stability, and compliance. Even for this very narrowly circumscribed pick-and-place style of manipulation, and with all these assumptions, we are missing a very big piece of the problem: path planning. How do we assure that MOVETO can go from the start to the goal without colliding with other blocks in the scene? This question is central to pick-and-place manipulation, and to manipulation in general.

Pick and place in practice

Before going on to the path planning problem, let's explore the pick-and-place idea as it occurs in manufacturing practice. First, consider the Sony SMART-Cell system introduced in chapter 1. It illustrates some elements that are widespread in industrial manipulation:

- Each gripper is designed to grasp a specific object.
- Parts feeders are designed to orient parts and bring them to known "initial" locations.
- Products are designed so that assembly is straightforward—most assembly operations are simple motions straight downward.
- Much of the path planning is simple: a DEPROACH moving the arm straight up to a safe height, a MOVETO to a point above the goal, and an APPROACH downward to the goal position.

Thus, many industrial assembly operations really do resemble pick-and-place manipulation. But there are also numerous exceptions to this generalization. In the Sony system, for example, the robot moves some already-placed parts, without grasping them, in preparation for placing the next part. The robot also uses an elaborate sequence of motions to thread a drive belt over a number of spindles.

Perhaps the main point to observe about industrial manipulation is that they have not eliminated the problems of grasp planning, motion planning, assembly planning, and so on. Rather, they have moved these planning problems off-line. For example, the parts feeding system cannot assume a known initial position for an object. It has to use a variety of mechanical tricks, sometimes augmented by sensing, to move the object to its "initial" position. Similarly, the assembly planning problem is transformed from: "What motion will correctly assemble these two given parts?" to the dual design-for-assembly problem: "What part design will be correctly assembled by a vertical motion?" The principles underlying these two problems are the same, and in particular, the mechanics of assembly are fundamental to both problems. Thus, later chapters will address the mechanics of assembly in detail. For the present, however, we return to the issue of path planning.

Kinematic Manipulation

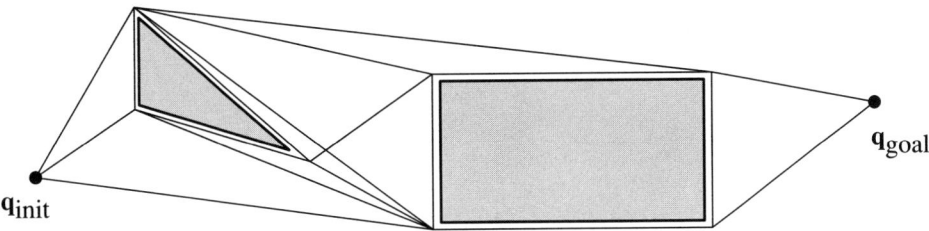

Figure 4.1
Visibility graph for a point robot with polygonal obstacles.

The configuration space transform

The first important concept in path planning is the *configuration space transform*. Recall that the *configuration space* is the space of configurations **q** for a given system. For example, if the system is a planar rigid object, the configuration space would be **SE**(2). Now, given fixed obstacles of known shape and location, we classify each point in configuration space as either *collides* if there is a collision, or *free* if there is not. The set of configurations labeled *collides* is the *configuration space obstacle*, or *C-space obstacle*.

Now any motion of the system corresponds to a curve **q**(t) in configuration space, so finding a collision-free motion means finding a curve **q**(t) that avoids the C-space obstacle. Thus we say that the configuration space transform reduces motion planning problems to the problem of finding a path for a single point in configuration space.

We illustrate the idea with some examples, then address the problem of path planning.

EXAMPLE 1: POINT IN PLANE

If the "robot" is a single moving point, the configuration space transform is trivial. The configuration of a point is just the location of the point itself, and the C-space obstacles are the same shape as the original obstacles. Figure 4.1 shows a simple example with polygonal obstacles, and also shows a simple planning technique: the *visibility graph*. We are given the obstacle shapes and locations, the initial robot pose q_{init}, and the final robot pose q_{goal}. The vertices of the graph are all obstacle vertices plus q_{init} and q_{goal}. Now for every pair of vertices, we include the line segment connecting them as a graph edge, if and only if that line segment is free of collisions. The shortest collision-free path from q_{init} to q_{goal} can be found by searching the visibility graph.

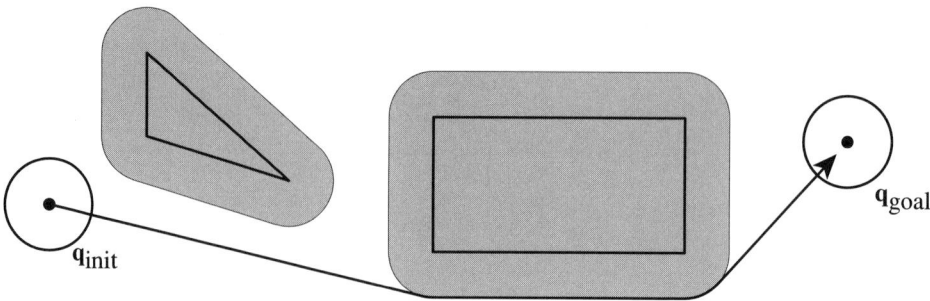

Figure 4.2
Configuration-space transform for a round planar robot.

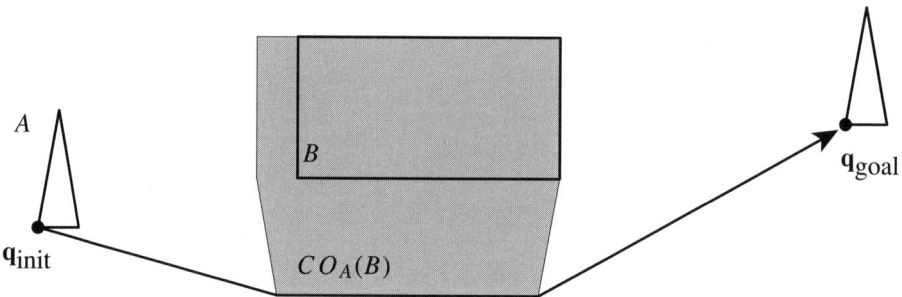

Figure 4.3
Configuration-space transform for a planar translating triangle.

EXAMPLE 2: ROUND MOBILE ROBOT IN PLANE

We can model a cylindrical mobile robot as a disk of radius r among a set of polygonal obstacles (figure 4.2). To avoid a collision, the center of the disk must be no closer than r to any obstacle. We can transform the problem to the case of a point robot, by expanding each obstacle by r.

This example illustrates the configuration-space transform: we obtain a moving point problem from a more complex problem by thinking in terms of *configuration space*, an approach which is very general. For the round mobile robot the idea is obvious: instead of thinking about motion of the disk, we think about motion of the center of the disk. The next two examples illustrate extensions of the idea.

Kinematic Manipulation

EXAMPLE 3: TRANSLATING POLYGON IN PLANE

Here we apply the configuration-space transform to the more interesting case of a translating polygon (figure 4.3). We represent the polygon's translation by the position $\mathbf{q} = (x, y)$ of a reference point fixed in the triangle. The transformed obstacles correspond to those values of (x, y) that would cause a collision.

The transformed obstacles are easy to construct using vector methods. We represent a convex polygonal obstacle B as the set of vectors giving points in B relative to some fixed origin. Let A be the moving convex polygon, represented as a set of vectors giving the points in A relative to the reference point location $\mathbf{q}$. Now if A and B overlap for some value of $\mathbf{q}$, then there must be some point included in both:

$$\text{collision} \leftrightarrow \exists_{\mathbf{a} \in A, \mathbf{b} \in B} \; \mathbf{a} + \mathbf{q} = \mathbf{b}$$

Let the *C-space obstacle* of B relative to A be defined:

$$CO_A(B) = \{\mathbf{q} \mid \exists_{\mathbf{a} \in A, \mathbf{b} \in B} \; \mathbf{a} + \mathbf{q} = \mathbf{b}\} \tag{4.1}$$
$$= \{\mathbf{b} - \mathbf{a} \mid \mathbf{a} \in A, \mathbf{b} \in B\} \tag{4.2}$$
$$= B \ominus A \tag{4.3}$$

where "$\ominus$" is sometimes called *Minkowski difference*. When the robot A and obstacle B are both convex obstacles, you can construct $CO_A(B)$ by starting with just the *vertices* of A and B, then taking the convex hull. (The convex hull of a set of points is just the smallest convex set containing the points. In the plane, it can be visualized as stretching a rubber band around all the points. In three-space it can be visualized as wrapping all the points up tightly in paper.) In other words

$$CO_A(B) = \text{conv}(\text{vert}(B) \ominus \text{vert}(A)) \tag{4.4}$$

which provides a very simple construction technique.

EXAMPLE 4: TRANSLATIONS AND ROTATIONS IN THE PLANE

If we also allow the triangle to rotate, the configuration space is three-dimensional (figure 4.4). We encode the triangle's configuration as $\mathbf{q} = (x, y, \theta)$, where (x, y) is the reference point coordinates and θ is the triangle's orientation. The simple techniques of the previous examples do not apply to rotations. I constructed the figure by sampling the triangle's orientation at 5-degree increments and applying the translation-only method of example 3.

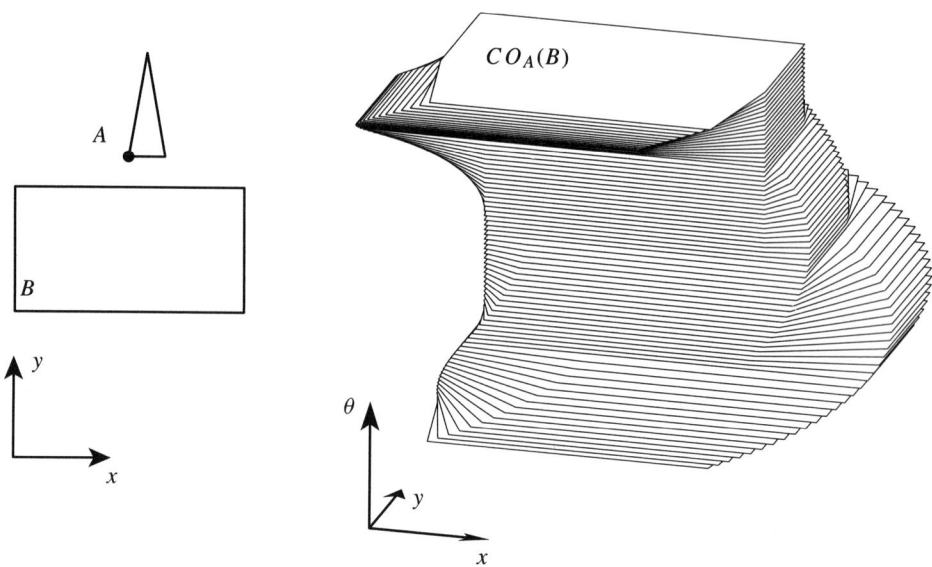

Figure 4.4
Configuration-space transform for a translating and rotating triangle.

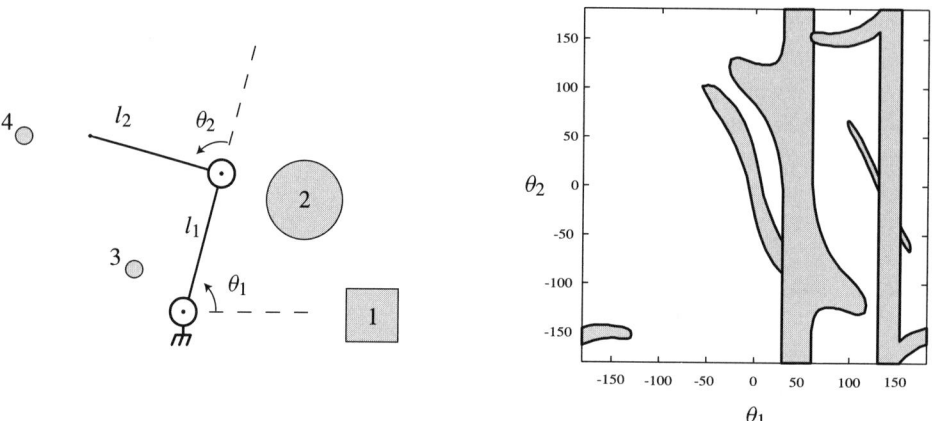

Figure 4.5
Configuration-space transform for a two link planar arm.

EXAMPLE 5: A TWO LINK PLANAR ARM

Given a mechanism with two revolute joints, the most natural encoding of configuration is by the two joint angles, $\mathbf{q} = (\theta_1, \theta_2)$. The topology of the configuration space is a torus, but I have flattened it (figure 4.5). The C-space obstacles were approximated by taking a discrete sampling of θ_2.

Path planning—heuristic search in discretized C-space

For the simple two-dimensional configuration spaces, we can search the visibility graph as in figure 4.1 to find paths, although there are drawbacks. For example, in some applications it is not wise to pass so near to the obstacles. In higher dimensional spaces, the visibility graph will not work, and must give way to other methods. In this section we briefly describe a method due to Barraquand and Latombe called BFP for "Best First Planner".

The approach uses a *potential field* defined on configuration space. For each configuration of the system, the potential is the sum of two terms: the distance to the goal, and a penalty for being close to obstacles. In particularly simple cases, path planning can be performed by just letting the system roll down the potential field to the goal, but that doesn't work in general—the system can get trapped in a local minimum.

Instead of just following the potential field, BFP uses the potential field as a heuristic to guide a search. We place a regular grid in configuration space, and perform a best-first search starting at $\mathbf{q}_{init}$. At each step in the search we examine the best node available, i.e. the node with the lowest potential. If that is the goal, then of course we are done. If not, then we add all neighboring nodes to the list for future consideration.

```
procedure BFP
    open ← {q_init}
    mark q_init visited
    while open ≠ {}
        q ← best( open )
        if q = q_goal return( success )
        for n ∈ neighbors( q )
            if n unvisited and not a collision
                insert( n, open )
                mark n visited
    return( failure )
```

The main data structure is a priority queue open which efficiently produces the best node it has been given, that is, the node with the lowest potential. Not shown in this code is the bookkeeping necessary to remember how the search arrived at the goal, and thereby return the desired path.

Note that BFP doesn't actually compute any configuration-space obstacles. It only needs to determine whether a given configuration is a collision, and to compute its

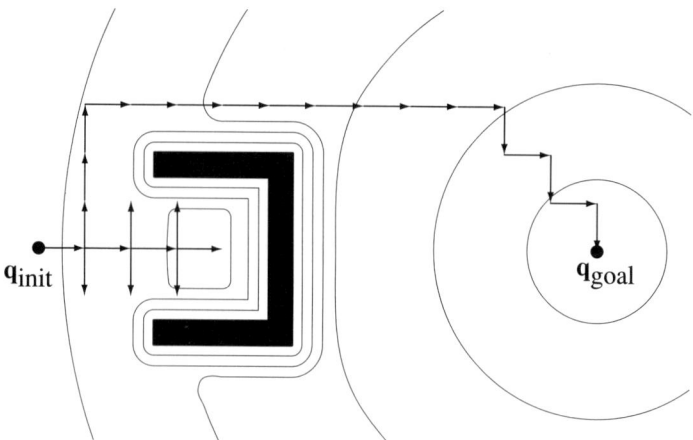

Figure 4.6
Using a potential field to guide search.

potential. The potential field includes a penalty for being close to an obstacle, which is approximated by calculating the proximity to obstacles of a few representative points on the system. We use the configuration-space idea while avoiding the daunting challenges of explicitly constructing a representation of the free configuration space.

A simple example is shown in figure 4.6. The search proceeds straight down the potential field until it hits the goal if we are lucky. If we hit a local minimum first, then the search has to visit all the nodes in the potential well before continuing.

A caveat: there is no planner yet devised that works well for all problems. I have chosen BFP because it illustrates the principles, and it has proven useful for some problems. However, when applying it to a new problem, you should be prepared to adapt it or discard it for some alternative approach.

4.2 Path planning for nonholonomic systems

The configuration space transform of the previous section can be applied to any holonomic system, at least in principle, simply by using the right configuration space. Even a constrained system may be addressed if it is holonomic, by working in the leaf of the foliation of configuration space.

For nonholonomic systems we have to work a little harder. We have to use actions that satisfy the velocity constraints, selected from a set of actions that generally violate the constraints. This section addresses path planning for such systems. The main example will be the unicycle introduced in section 2.5. Later (section 7.3) we will also apply the techniques to a pushing problem.

In theory, there is a cute trick that avoids the complications of nonholonomic systems. First, compute the involutive closure of the system. That defines a holonomic system for which we can plan a path using, for instance, our best-first planner of the previous section. Now the resulting path will not be consistent with the constraints, but our robot can follow the path approximately using Lie bracket motions, assuming that the constraints are Pfaffian. Thus, for example, we can plan a path for an automobile as if it were a free-flying robot, and then approximate this path using lots of tiny parallel-parking maneuvers.

Clearly, we need a more practical approach. Here we apply a method similar in spirit to BFP, and which was also developed by Barraquand and Latombe. Again it is forward-chaining best first search, but we use a discretized space of actions to generate the graph to search. As we search, we prune collisions. We also prune nodes that are close to previously generated nodes. That is accomplished by keeping a grid in configuration space, and marking the grid node closest to each configuration we generate. We will call the procedure NHP for NonHolonomic Planner. The parameter δt is some suitably small increment of time. The routine grid(q) returns the closest grid node to q. The parameter actions is a finite set of actions. The routine int(q, a, δt) integrates the system forward from q using action a for time δt and returns the resulting configuration.

```
procedure NHP
    open ← {q_init}
    mark grid( q_init ) visited
    while open ≠ {}
        q ← best( open )
        if q ≈ q_goal return( success )
        for a ∈ actions
            n ← int( q, a, δt )
            if n not a collision and grid( n ) not visited
                insert( n, open )
                mark grid( n ) visited
    return( failure )
```

Several choices remain open. The discrete set actions must be chosen so that we do not unduly restrict the possible motions of the system. However, the number of actions must be kept small, because the running time is exponential in the number of actions. In the case of the unicycle, actions might include forward and reverse rolling, and left and right turning. That excludes motions that combine simultaneous rolling and turning, which may or may not seem acceptable. Another open choice is the cost function that

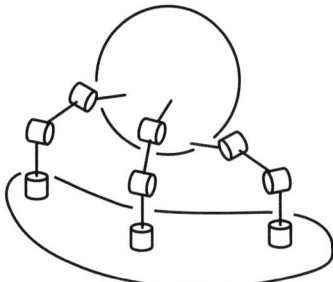 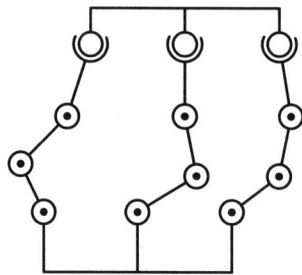

Figure 4.7
An object held by three pointy fingers, and a kinematic model of the grasp.

determines which node is best. We might choose a cost function corresponding to the length of the best path found to that node, and include a penalty for each switch of actions. For the unicycle, that choice would tend to give short and smooth paths, avoiding paths with frequent alternation between rolling and turning.

Allow me to repeat the caveat from the previous section: there is no planner yet devised that works well for all problems. I have chosen NHP because it illustrates the principles, and it has proven useful for some problems. However, when applying it to a new problem, you should expect to adapt it or discard it for some alternative approach.

4.3 Kinematic models of contact

A recurring theme in robotic manipulation is how to model contact between fingers and objects. In some cases a very simple kinematic approach can be taken—we can model the contact as a joint in a kinematic linkage. For example, if we model point finger contacts as spherical joints, then the grasp of figure 4.7 might be modeled as a spatial linkage with nine revolute joints and three spherical joints. As long as the fingers do not slip or break contact altogether, we can analyze the task using the tools of kinematic mechanisms, Grübler's formula in particular.

What is the nature of the joint formed where a finger makes contact with an object? It depends on the shape, stiffness, and frictional characteristics of the finger and the object. In general the interaction between finger and task is quite complex, but simple cases can be modeled using a taxonomy developed by Salisbury (figure 4.8).

Kinematic Manipulation

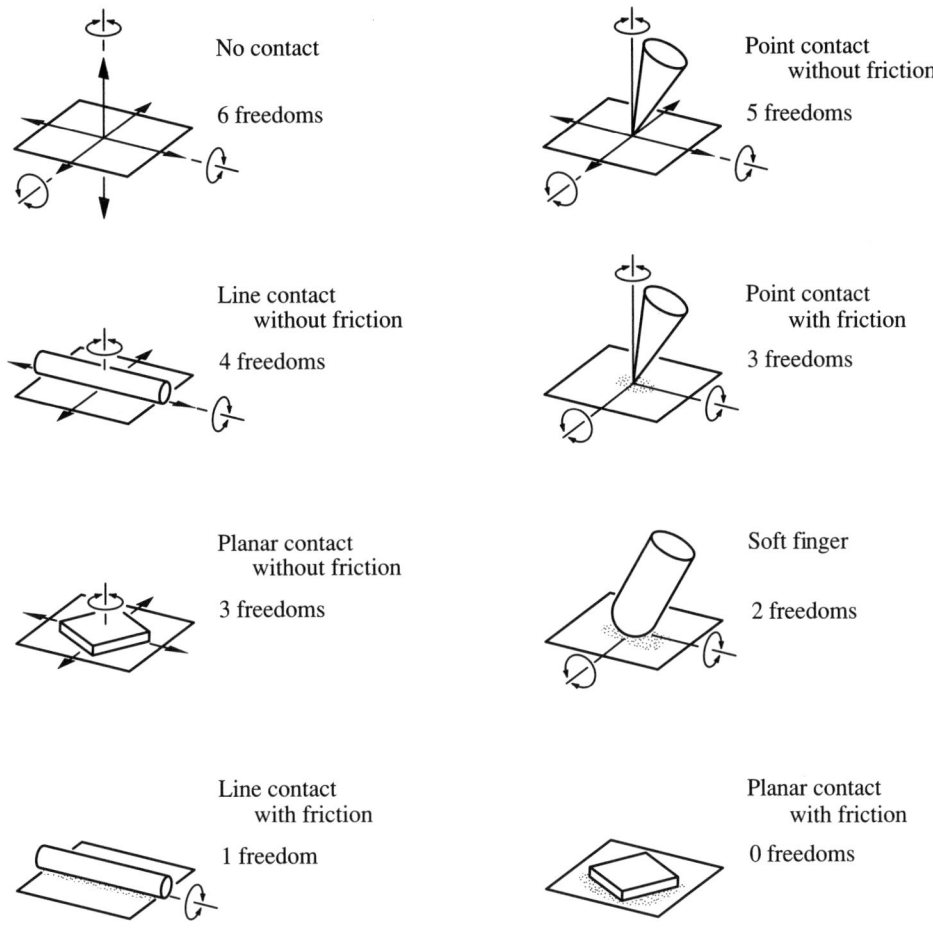

Figure 4.8
Salisbury's table of contact types.

Number synthesis of grasp

Design of mechanisms is often divided into three stages: *number synthesis* which means counting freedoms and constraints; *type synthesis* which means choosing the type of motion allowed at the joints; and *dimensional synthesis* which means choosing specific dimensions for the links.

We can use Grübler's formula to address number synthesis of grasp, following Salisbury (1982; 1985). Consider the grasp of figure 4.7. If we model the finger contacts as type "point contact with friction" then for each of three finger contacts we have three freedoms. We also have nine finger joints, each with one freedom. There are two loops. Grübler's formula gives a mobility of six. While that is promising, the mobility of the entire system does not really tell us everything we need to know. Salisbury suggests four measures:

M Mobility of the entire system with the finger joints free.
M' Mobility of the entire system, with the finger joints locked.
C Connectivity of the object relative to a fixed palm, with the finger joints free.
C' Connectivity of the object relative to a fixed palm, with the finger joints locked.

We then accept a design with $C = 6$ and $C' \leq 0$. When the finger joints are free, we want connectivity of six so that the object can make general motions. When the finger joints are locked, we want connectivity of zero so that the hand can immobilize the object. The example of figure 4.7 satisfies these criteria with $C = 6$ and $C' = 0$.

4.4 Bibliographic notes

For path planning the primary reference is Latombe's text (1991). The configuration space transform originated with (Lozano-Pérez and Wesley, 1979). Potential fields were introduced by (Khatib, 1980, 1986). The best-first planner appears in (Barraquand and Latombe, 1991). The nonholonomic planner appears in (Barraquand and Latombe, 1993). The application of nonlinear geometric control theory to nonholonomic robotic systems originated with (Li and Canny, 1990).

The material on kinematic models of grasp and number synthesis of grasp are taken from Salisbury's PhD thesis (1982; 1985). Since Salisbury's seminal work, interest in manipulation in the hand (frequently called "dexterous manipulation") has grown enormously. Rolling contact between fingers and object was analyzed by Montana (1988) and forms the basis for work on planning the motions of fingers. See Murray, Li, and Sastry's text (1994) for an introduction and (Okamura, Smaby, and Cutkosky, 2000) for a survey. There have also been numerous advances in the modeling of contact; see (Bicchi and Kumar, 2000) for a survey.

Exercises

Exercise 4.1: Let A be a mobile unit-edge square robot which can translate but not rotate in the plane (figure 4.9). Let B be a unit-edge equilateral triangle obstacle. Use the simple procedure suggested by equation 4.4 to construct $CO_A(B)$:

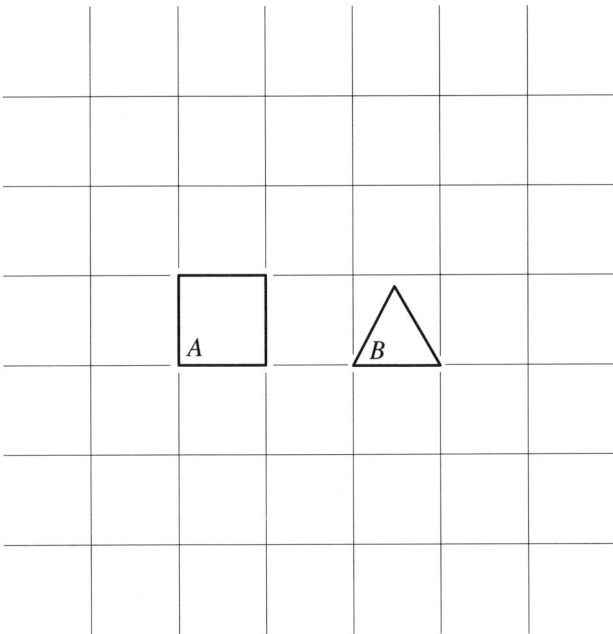

Figure 4.9
For exercise 4.1.

- Plot the vertices of the obstacle B.
- Choose a reference point **q** on A and construct $\ominus A$ by reflecting every vertex through **q**.
- Plot the vertices of $B \ominus A$: for each vertex of B plot a copy of $\ominus A$ with **q** coincident with that vertex.
- Construct the convex hull.

Exercise 4.2: Use the same procedure as exercise 4.1 to construct the C-space obstacle for figure 4.10. The robot is a unit-edge equilateral triangle, which can translate but not rotate in the plane.

Exercise 4.3: We can adapt the procedure of exercise 4.1 for concave polygonal obstacles as follows: we divide the concave object into convex polygons. We construct C-space obstacles for each convex sub-obstacle. The union gives the total C-space obstacle. Use this method to construct the C-space obstacle of figure 4.11. Here the robot is a unit-edge equilateral triangle that can translate but not rotate in the plane.

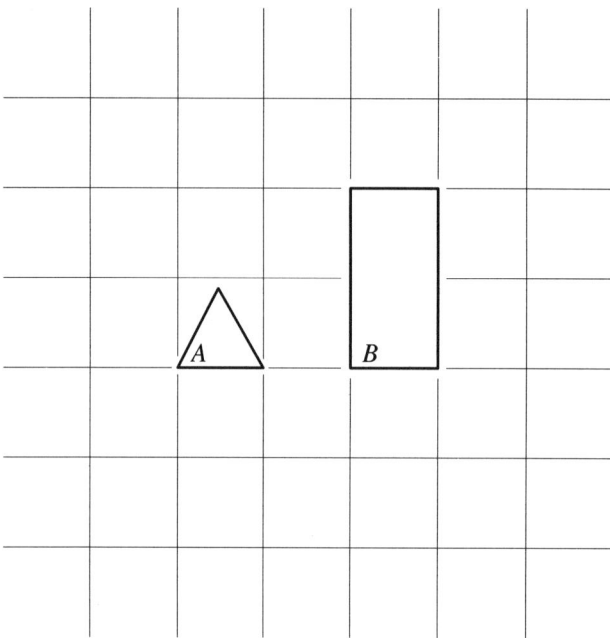

Figure 4.10
For exercise 4.2.

Exercise 4.4: Consider the C-space obstacle of figure 4.4. Each surface facet corresponds to a vertex-edge pair or an edge-vertex pair, that is, either a vertex of A and an edge of B; or an edge of A and a vertex of B. Pick a vertex-edge type surface visible on the C-space obstacle in the figure, and identify the corresponding features on A and B. Do the same for an edge-vertex type surface.

Exercise 4.5: Label the C-space obstacles of figure 4.5 with the number of the corresponding actual obstacle.

Exercise 4.6: How many connected components does the free space have in figure 4.5? Remember that the true topology of this C-space is a torus, obtained by gluing the top edge to the bottom edge, and gluing the left edge to the right edge.

Exercise 4.7: Although the best-first planner BFP can escape from potential wells, it still does not generate optimal plans. Design a path-planning problem for which BFP produces an absurdly long path.

Kinematic Manipulation

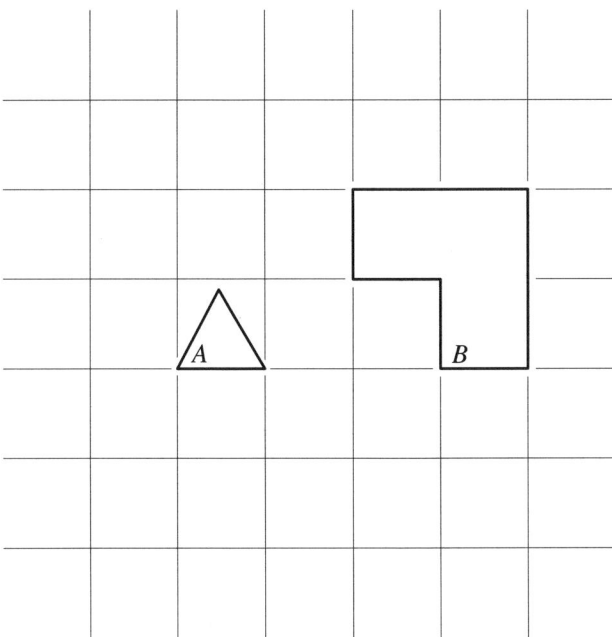

Figure 4.11
For exercise 4.3.

Exercise 4.8: How would you apply the nonholonomic planner NHP to the refrigerator walking problem of exercise 2.4? What finite set of actions would you choose? What value of δt? What resolution of the configuration-space grid? What cost function?

Exercise 4.9: Implement NHP and try it out on the refrigerator walking problem.

Exercise 4.10: For each of Salisbury's criteria, design a spatial grasp that fails to meet the criterion, i.e. such that $C < 6$ or $C' > 0$.

Exercise 4.11: Calculate C and C' for the grasp of figure 4.7 using contacts of type "point contact without friction".

Exercise 4.12: Calculate C and C' for the grasp of figure 4.7 using contacts of type "soft finger".

Exercise 4.13: Construct a taxonomy of contact types for planar grasps, similar to that of figure 4.8.

Exercise 4.14: Design a planar hand and grasp that meets Salisbury's criteria, achieving connectivities $C = 3$ and $C' \leq 0$. Design one that fails to meet the first criterion $C = 3$. Design one that fails to meet the second criterion $C' \leq 0$. In each case don't forget to state which model of contact you assume.

5 Rigid Body Statics

Statics is the study of force—how to characterize a force, how to characterize a set of forces, and how forces are distributed in structures. This chapter presents the basic properties of forces acting on rigid bodies, and then develops methods for representing and analyzing systems of forces.

5.1 Forces acting on rigid bodies

This section addresses the problem of how to characterize forces acting on a rigid body. To begin, we adopt axioms that apply to static forces acting on particles. We hypothesize:

1. A force applied to a particle can be described as a vector.
2. The motion of a particle is determined by the vector sum of the forces applied to the particle.
3. In particular, a particle will remain at rest only if the total force acting on it is zero.

Now we define the *moment of force about a line*, or *torque about a line*. Let l be some line with direction $\hat{\mathbf{l}}$, passing through the origin. Suppose a force $\mathbf{f}$ acts at $\mathbf{x}$. Then the moment of $\mathbf{f}$ about l is defined to be

$$n_l = \hat{\mathbf{l}} \cdot (\mathbf{x} \times \mathbf{f}) \tag{5.1}$$

We can also define the *moment of force* or *torque* about a point $\mathbf{O}$ to be

$$\mathbf{n}_O = (\mathbf{x} - \mathbf{O}) \times \mathbf{f} \tag{5.2}$$

If the origin is $\mathbf{O}$ this reduces to

$$\mathbf{n} = \mathbf{x} \times \mathbf{f} \tag{5.3}$$

If $\mathbf{n}$ gives the moment about the origin, and n_l gives the moment about the line l through the origin, then note that

$$n_l = \hat{\mathbf{l}} \cdot \mathbf{n} \tag{5.4}$$

Now suppose we have a rigid body. Forces acting on the rigid body can be divided into two classes: *internal forces* are those forces acting between particles of the body; and *external forces* are those forces acting from outside the body. We define the *total force* $\mathbf{F}$ acting on the body to be the sum of all the external forces, and the *total moment* (or total

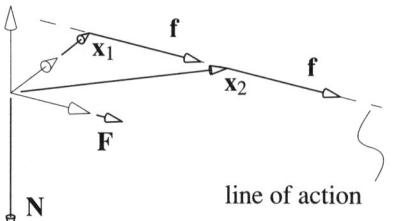

Figure 5.1
A force may be applied at any point on its line of action without changing its effect on a rigid body.

torque) **N** to be the sum of all the corresponding moments about the origin:

$$\mathbf{F} = \sum \mathbf{f}_i \qquad (5.5)$$
$$\mathbf{N} = \sum \mathbf{x}_i \times \mathbf{f}_i \qquad (5.6)$$

It is a consequence of Newton's laws that the effect of any system of forces, acting on a single rigid body, is completely determined by the total force **F** and the total moment **N**. (This point is considered in greater detail in chapter 8.) Any two systems of forces giving identical **F**, **N**, are said to be equivalent. If there is a single force with the same **F**, **N**, it is called the *resultant*.

Line of action. When a force is applied to a particle in three dimensions, that force is completely characterized as a three-dimensional vector. But when a force is applied to a rigid body, we have to consider at which *point* the force is applied. Consider a force **f** applied at some point **x**. The total force **F** is just the given force: **F** = **f** But the total moment **N** depends on the point of application: **N** = **x** × **f**. However, **N** is not changed if the force is moved along the line in which it lies—its *line of action*. In other words, if $(\mathbf{x}_2 - \mathbf{x}_1) \parallel \mathbf{f}$, then it does not matter whether the force is applied at $\mathbf{x}_1$ or at $\mathbf{x}_2$. Since **f** can vary freely along a line, it is sometimes called a *line vector*. Vectors that are fixed at a single point are sometimes called *bound vectors*, and ordinary vectors are called *free vectors* when it is necessary to distinguish them.

Resultant of two forces on intersecting lines of action. Suppose two forces $\mathbf{f}_1$ and $\mathbf{f}_2$, acting on L_1 and L_2 respectively, are applied to some rigid body. If the lines of action intersect, then it is simple to construct the resultant—the single force equivalent to the original system of two forces. We can slide both $\mathbf{f}_1$ and $\mathbf{f}_2$ along their respective lines of action to the intersection. Now we have two forces acting at a common point. The resultant force is the vector sum $\mathbf{f}_1 + \mathbf{f}_2$, acting at the intersection.

Rigid Body Statics

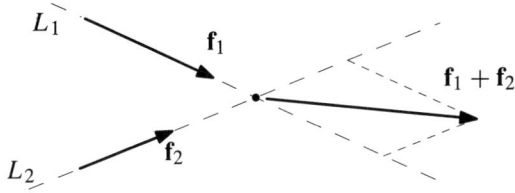

Figure 5.2
The resultant of two forces with intersecting lines of action.

Change of reference. Suppose some system of forces yields a total force and moment $\mathbf{F}_Q, \mathbf{N}_Q$, taken with respect to the point Q. What are the force and moment $\mathbf{F}_R, \mathbf{N}_R$, for a different reference point R? Write the total force and total moment for each choice of reference point:

$$\mathbf{F}_R = \sum \mathbf{f}_i \tag{5.7}$$

$$\mathbf{F}_Q = \sum \mathbf{f}_i \tag{5.8}$$

$$\mathbf{N}_R = \sum (\mathbf{x}_i - \mathbf{R}) \times \mathbf{f}_i \tag{5.9}$$

$$\mathbf{N}_Q = \sum (\mathbf{x}_i - \mathbf{Q}) \times \mathbf{f}_i \tag{5.10}$$

From which it follows

$$\mathbf{F}_R = \mathbf{F}_Q \tag{5.11}$$

and

$$\mathbf{N}_R - \mathbf{N}_Q = \sum (\mathbf{Q} - \mathbf{R}) \times \mathbf{f}_i \tag{5.12}$$

which gives

$$\mathbf{N}_R = \mathbf{N}_Q + (\mathbf{Q} - \mathbf{R}) \times \mathbf{F} \tag{5.13}$$

DEFINITION 5.1: A **couple** is a system of forces whose total force $\mathbf{F} = \sum \mathbf{f}_i$ is zero. Notice that the moment $\mathbf{N}$ of a couple is independent of reference point.

For an arbitrary couple it is trivial to construct an equivalent system of just two forces, which might explain the origin of the name. However, there is no equivalent system of just one force. That is, a couple of non-zero torque does not have a resultant. That means that not every system of forces can be characterized by a resultant.

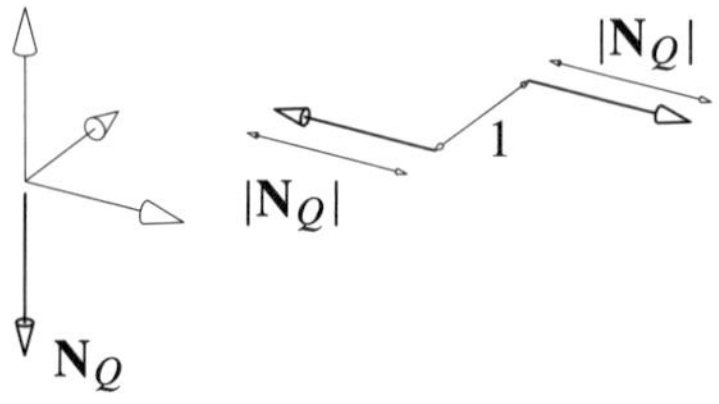

Figure 5.3
Construction for proof of theorem 5.2.

Equivalent systems of forces

We saw in the previous section that we can characterize a system of forces by its total force and moment, at least as far as its effect on a rigid body. We also saw that a resultant is not a general way of characterizing systems of forces, because some systems, couples for example, do not have resultants. There are other ways of characterizing a system of forces that are more general. In particular, we will define a *wrench*, which is analogous to the *twist* used to characterize rigid body motion. First we must tend to some preliminaries.

THEOREM 5.1: For any reference point Q, any system of forces is equivalent to a single force through Q, plus a couple.

Proof Let $\mathbf{F}$ be the total force, let $\mathbf{N}_Q$ be the total moment at Q. If we apply a single force $\mathbf{F}$ at Q, and construct a couple with moment $\mathbf{N}_Q$, then the total force and moment will be $\mathbf{F}$ and $\mathbf{N}_Q$, as required. ∎

THEOREM 5.2: Every system of forces is equivalent to a system of just two forces.

Proof It was remarked earlier that for any couple there is an equivalent system of two forces, and that couples can be moved rigidly without affecting their force or torque. So, take the construction of figure 5.3, using only two forces in the construction of the couple. We now have three forces: two to construct the couple, and one passing through Q. Move the couple so that one of its two forces passes through Q. Then replace the two forces at Q with their vector sum. Thus the three forces have been reduced to two. ∎

THEOREM 5.3: A system consisting of a single non-zero force plus a couple in the same plane, i.e. a torque vector perpendicular to the force, has a resultant.

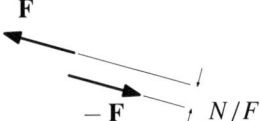

Figure 5.4
Construction for proof of theorem 5.3.

Proof Let **F** be the force, acting at P. Let **N** be the moment of the couple. Construct an equivalent couple as in figure 5.4 and translate it so that $-\mathbf{F}$ is applied at P. This will cancel the original **F**, leaving one resultant force. ∎

THEOREM 5.4 POINSOT'S THEOREM: Every system of forces is equivalent to a single force, plus a couple with moment parallel to the force.

Proof Let **F** and **N** be the force and moment, respectively, of a given system of forces. Decompose the moment into two components: $\mathbf{N}_\parallel$ parallel to **F**, and $\mathbf{N}_\perp$ perpendicular to **F**. By theorem 5.3 we can replace **F** and $\mathbf{N}_\perp$ by a single force $\mathbf{F}'$ parallel to **F**. Now we construct a couple with moment $\mathbf{N}_\parallel$ to obtain the desired result: a force and a couple with moment parallel to the force. ∎

Poinsot's theorem is analogous to Chasles's theorem (theorem 2.7). And, like Chasles's theorem, it can be phrased in terms of screws. First we define a *wrench*.

DEFINITION 5.2: A **wrench** is a screw plus a scalar magnitude, giving a force along the screw axis plus a moment about the screw axis. The force magnitude is the wrench magnitude, and the moment is the twist magnitude times the pitch. Thus the pitch is the ratio of moment to force.

Using the language of screws, Poinsot's theorem is succinctly stated: every system of rigid body forces reduces to a wrench along some screw.

We can also extend screw coordinates to include wrenches. Let f be the magnitude of the force acting along a line l, and let n be the magnitude of the moment about l. The magnitude of the twist is the magnitude of the force f. Starting from the definition of screw coordinates based on Plücker coordinates, we can write the screw coordinates of the wrench

$$\mathbf{w} = f\mathbf{q} \tag{5.14}$$
$$\mathbf{w}_0 = f\mathbf{q}_0 + fp\mathbf{q} \tag{5.15}$$

where $(\mathbf{q}, \mathbf{q}_0)$ are the normalized Plücker coordinates of the wrench axis l, and p is the pitch, which is defined to be

$$p = n/f \tag{5.16}$$

Let $\mathbf{r}$ be some point on the wrench axis, so we obtain

$$\mathbf{q}_0 = \mathbf{r} \times \mathbf{q} \tag{5.17}$$

Then by substituting equations 5.16 and 5.17 into equations 5.14 and 5.15 we can write

$$\mathbf{w} = \mathbf{f} \tag{5.18}$$
$$\mathbf{w}_0 = \mathbf{r} \times \mathbf{f} + \mathbf{n} \tag{5.19}$$

By comparing with equation 5.13 we can write

$$\mathbf{w} = \mathbf{f} \tag{5.20}$$
$$\mathbf{w}_0 = \mathbf{n}_0 \tag{5.21}$$

where $\mathbf{n}_0$ is just the moment of force at the origin. Thus we find that screw coordinates of a wrench are actually a familiar representation $(\mathbf{f}, \mathbf{n}_0)$. This yields a vector space, so that we can scale wrenches or add wrenches, just as with differential twists. For a wrench in the x-y plane, the f_z, n_{0x}, and n_{0y} terms are all zero, so planar wrenches can be written $(f_x, f_y, 0, 0, 0, n_{0z})$, or more simply as (f_x, f_y, n).

The reciprocal product of a differential twist and a wrench is meaningful and useful. Using screw coordinates:

$$(\boldsymbol{\omega}, \mathbf{v}_0) * (\mathbf{f}, \mathbf{n}_0) = \mathbf{f} \cdot \mathbf{v}_0 + \mathbf{n}_0 \cdot \boldsymbol{\omega} \tag{5.22}$$

which is the power produced by the wrench $(\mathbf{f}, \mathbf{n}_0)$ and differential twist $(\boldsymbol{\omega}, \mathbf{v}_0)$. Thus we can immediately observe that a differential twist is reciprocal to a wrench if and only if no power would be produced. In section 3.3, we developed a first order analysis of kinematic constraint using the reciprocal product between a velocity twist and a constraint screw. In section 5.3, we will use reciprocal product between a velocity twist and a wrench instead, but the result is the same.

Undoubtedly, the reader has observed that some conventions for wrench coordinates seem to be reversed from the conventions taken for twist coordinates. In particular, pitch $p = n/f$ has the angular component over the linear component, and the screw coordinates $(\mathbf{f}, \mathbf{n}_0)$ have the linear component before the angular component. Both of these are the reverse of the conventions for twist coordinates. This is not a peculiarity of our conventions; it reflects a deeper fact that is fundamental to the dual relationship of motion and force. For example, in comparing Chasles's and Poinsot's theorems, we find that an axis of rotation is

Rigid Body Statics

analogous to a line of force. Let us summarize some points of comparison between motion and force:

Motion	Force
A zero-pitch twist is a pure rotation.	A zero-pitch wrench is a pure force.
For a pure translation, the direction of the axis is determined, but the location is not.	For a pure moment, the direction of the axis is determined, but the location is not.
A differential translation is equivalent to a rotation about an axis at infinity.	A moment of force is equivalent to a force along a line at infinity.
In the plane, any motion can be described as a rotation about some point, possibly at infinity.	In the plane, any system of forces reduces to a single force, possibly at infinity.

5.2 Polyhedral convex cones

Polyhedral convex cones occur naturally when describing the sets of wrenches and twists that characterize rigid body contact. This section develops the essential properties of cones in the n-dimensional vector space $\mathbf{R}^n$. The results can be applied to wrench space or velocity twist space.

Let $\mathbf{v}$ be any non-zero vector in $\mathbf{R}^n$. Then the set of vectors

$$\{k\mathbf{v} \mid k \geq 0\} \tag{5.23}$$

describes a ray of $\mathbf{R}^n$ (figure 5.5a). Similarly, if $\mathbf{v}_1$ and $\mathbf{v}_2$ are two non-zero and non-parallel vectors in $\mathbf{R}^n$, then the set of vectors

$$\{k_1\mathbf{v}_1 + k_2\mathbf{v}_2 \mid k_1, k_2 \geq 0\} \tag{5.24}$$

describes a planar cone (figure 5.5c). We can generalize to an arbitrary number of vectors by defining the *positive linear span* of a set of vectors $\{\mathbf{v}_i\}$:

$$\text{pos}(\{\mathbf{v}_i\}) = \{\sum k_i \mathbf{v}_i \mid k_i \geq 0\} \tag{5.25}$$

We will take the positive linear span of the empty set to be just the origin. Contrast the positive linear span with two related constructs: the *linear span*

$$\text{lin}(\{\mathbf{v}_i\}) = \{\sum k_i \mathbf{v}_i \mid k_i \in \mathbf{R}\} \tag{5.26}$$

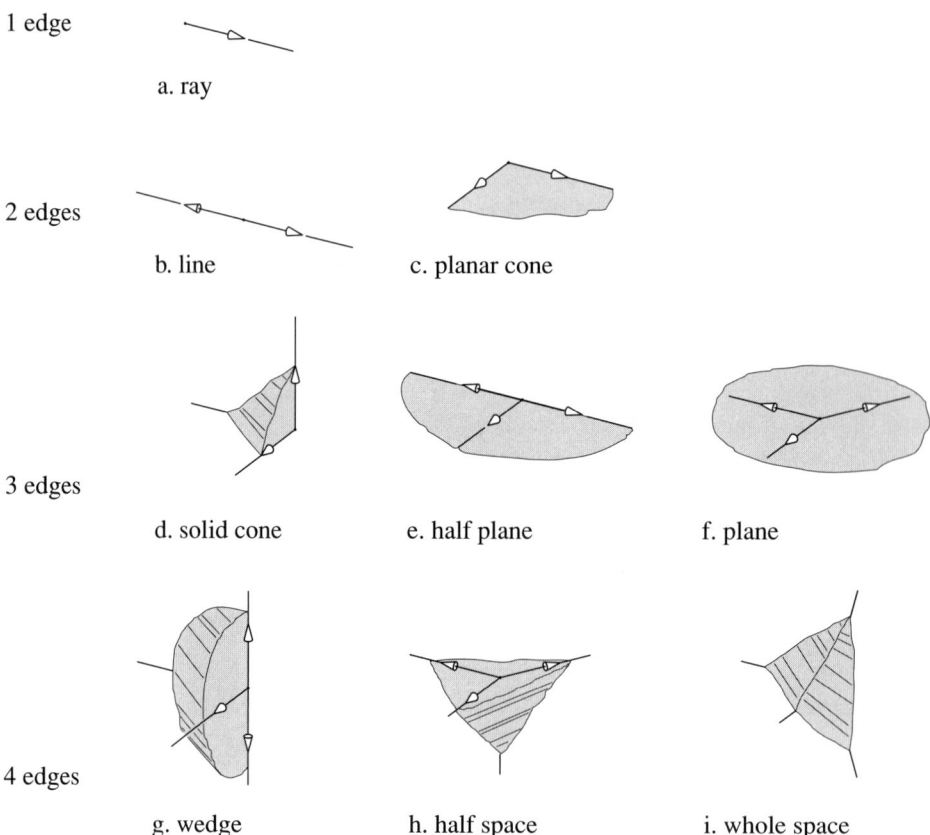

Figure 5.5
Polyhedral convex cones.

and the *convex hull*

$$\text{conv}(\{\mathbf{v}_i\}) = \{\sum k_i \mathbf{v}_i \mid k_i \geq 0, \sum k_i = 1\} \tag{5.27}$$

By taking positive linear spans it is possible to construct rays, lines, half-planes, and so on (figure 5.5). Any set of vectors constructed in this fashion is a *polyhedral convex cone*. Of particular interest is the case where the positive linear span is the entire space: $\text{pos}(\{\mathbf{v}_i\}) = \mathbf{R}^n$. We offer two theorems without proof.

Rigid Body Statics

THEOREM 5.5: A set of vectors $\{\mathbf{v}_i\}$ positively spans the entire space $\mathbf{R}^n$ if and only if the origin is in the interior of the convex hull:

$$\text{pos}(\{\mathbf{v}_i\}) = \mathbf{R}^n \leftrightarrow \mathbf{0} \in \text{int}(\text{conv}(\{\mathbf{v}_i\})) \tag{5.28}$$

THEOREM 5.6: It takes at least $n + 1$ vectors to positively span $\mathbf{R}^n$.

Thus in $\mathbf{R}^3$, it takes at least 4 vectors to positively span the space, which should be evident from studying figure 5.5.

Representing cones; supplementary cones

There are two ways to represent a polyhedral convex cone: by its edges and by its faces. To represent a cone by its edges, we already have the relevant operator: positive linear span. Given a set of edges $\{\mathbf{e}_i\}$, the cone is given by $\text{pos}(\{\mathbf{e}_i\})$.

To represent a cone by a set of faces, we begin by representing a planar half-space by its inward pointing normal vector $\mathbf{n}$. Define the positive half space determined by a vector $\mathbf{n}$ to be

$$\text{half}(\mathbf{n}) = \{\mathbf{v} \mid \mathbf{n} \cdot \mathbf{v} \geq 0\} \tag{5.29}$$

(Here we use dot product, but when working with twists and wrenches we will use reciprocal product.)

Then we can represent a polyhedral convex cone as the intersection of half-spaces:

$$\cap \{\text{half}(\mathbf{n}_i)\} \tag{5.30}$$

There is one last definition of interest. Suppose we are given a cone's face normals, but we treat them as edges and take their positive linear span. Or, suppose we are given the edges and treat them as normals. Then we will get a different cone, called the *supplementary cone* (figure 5.6).

DEFINITION 5.3: Given a polyhedral convex cone V:

$$V = \text{pos}(\{\mathbf{e}_i\}) = \cap \{\text{half}(\mathbf{n}_i)\} \tag{5.31}$$

we define the **supplementary cone** to be

$$\text{supp}(V) = \text{pos}(\{\mathbf{n}_i\}) = \cap \{\text{half}(\mathbf{e}_i)\} \tag{5.32}$$

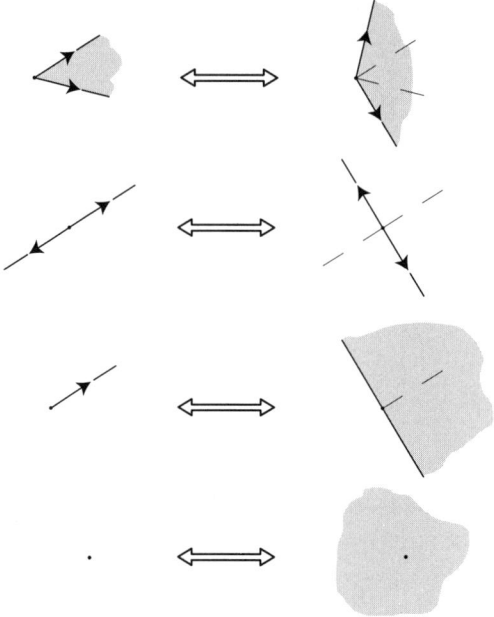

Figure 5.6
Supplementary cones.

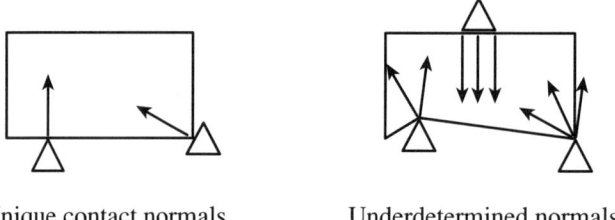

Unique contact normals   Underdetermined normals

Figure 5.7
Analysis of contact is sometimes problematic when a unique contact normal does not exist.

5.3 Contact wrenches and wrench cones

Polyhedral convex cones are useful for reasoning about frictionless contact on rigid bodies. For the present we will consider a single frictionless contact with a unique contact normal (figure 5.7), and we will assume that a frictionless contact can apply a force of arbitrary non-negative magnitude along the inward-pointing contact normal.

Rigid Body Statics

Let $\mathbf{w} = (\mathbf{c}, \mathbf{c}_0)$ be the vector of normalized screw coordinates of the inward-pointing contact normal. Then by assumption the contact wrench must be of the form

$$k\mathbf{w}, k \geq 0 \tag{5.33}$$

In other words, the wrench must be in the ray given by the positive linear span pos($\mathbf{w}$).

Now suppose we have two frictionless contacts $\mathbf{w}_1$ and $\mathbf{w}_2$. The total wrench acting on the object is obtained by summing the contributions from $\mathbf{w}_1$ and $\mathbf{w}_2$:

$$k_1\mathbf{w}_1 + k_2\mathbf{w}_2; k_1, k_2 \geq 0 \tag{5.34}$$

so that the set of all possible wrenches due to two frictionless contacts would be given by the positive linear span pos($\{\mathbf{w}_1, \mathbf{w}_2\}$).

Generalizing to an arbitrary number of contacts, it follows immediately from our assumptions on frictionless contact that:

THEOREM 5.7: If a set of frictionless contacts on a rigid body is described by the contact normals $\mathbf{w}_i = (\mathbf{c}_i, \mathbf{c}_{0i})$ then the set of all possible wrenches is given by the positive linear span pos($\{\mathbf{w}_i\}$).

Thus we see that every set of frictionless contacts on a rigid body corresponds to a polyhedral convex cone in wrench space. Some important results follow immediately. First a definition:

DEFINITION 5.4: **Force closure** means that the set of possible wrenches exhausts all of wrench space.

Force closure is often used to characterize a grasp or a fixture intended to immobilize an object. It follows from theorem 5.6 that a frictionless force closure grasp requires at least 7 fingers. Or, since planar wrench space is only three-dimensional, a frictionless force closure grasp in the plane requires at least 4 fingers.

EXAMPLE: FORCE CLOSURE GRASP

Consider a rigid square in the plane grasped by four fixed frictionless fingers (figure 5.8). First we need to write screw coordinates for the four wrenches. Take contact 1 as an example. We imagine a unit force in the y direction, so $(f_{1x}, f_{1y}) = (0, 1)$. It is applied at the point $(p_{1x}, p_{1y}) = (1, -1)$. The corresponding torque is $\tau = 1$, which may be calculated by taking the cross product $\mathbf{p}_1 \times \mathbf{f}_1$, or in such a simple case it is easily obtained by inspection of the drawing. So the screw coordinates of contact normal 1 are

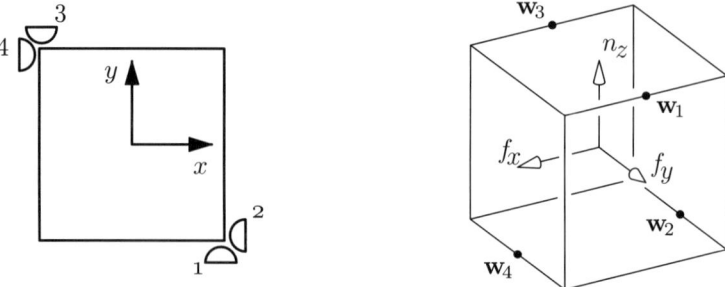

Figure 5.8
The origin of wrench space is in the interior of the convex hull of the four wrenches, so the grasp is force closure.

$(f_{1x}, f_{1y}, \tau_1) = (0, 1, 1)$. The other contacts are handled in similar fashion, yielding:

$$w_1 = \begin{pmatrix} 0 \\ 1 \\ 1 \end{pmatrix} \tag{5.35}$$

$$w_2 = \begin{pmatrix} -1 \\ 0 \\ -1 \end{pmatrix} \tag{5.36}$$

$$w_3 = \begin{pmatrix} 0 \\ -1 \\ 1 \end{pmatrix} \tag{5.37}$$

$$w_4 = \begin{pmatrix} 1 \\ 0 \\ -1 \end{pmatrix} \tag{5.38}$$

If we plot these four wrenches in wrench space, we note that their configuration determines a tetrahedron with the origin in the middle. By theorem 5.5 the grasp is force closure.

5.4 Cones in velocity twist space

A twist describing a finite displacement cannot be regarded as a vector, so our construction of polyhedral convex cones using positive linear span does not apply to finite displacement twists.

Rigid Body Statics

But a velocity twist in screw coordinates has the form $\mathbf{t} = (\boldsymbol{\omega}, \mathbf{v}_0)$, which can treated as a vector. Positive linear span and polyhedral convex cones are applicable in velocity twist space.

Let $\{\mathbf{w}_i\}$ be a set of contact normals. Let W be the corresponding set of possible contact wrenches, $W = \text{pos}(\{\mathbf{w}_i\})$. Recall our first-order kinematic analysis of kinematic constraint: any feasible velocity twist must be reciprocal or repelling to the contact screws. (See section 3.3.) Let T be the set of feasible (to first order) velocity twists. For every $\mathbf{t}$ in T, for every contact screw $\mathbf{w}_i$, $\mathbf{t} * \mathbf{w}_i \geq 0$. Each contact screw determines a half-space in velocity twist space, and T is the intersection of the half-spaces:

$$T = \bigcap \{\text{half}(\mathbf{w}_i)\} \tag{5.39}$$

showing that T is a polyhedral convex cone in twist space. Using the language of cones, we can say it succinctly: the cone of reciprocal or repelling twists is supplementary to the cone of contact wrenches.

As an example, consider Example 1 of section 3.3, where a cube is constrained by six contact screws. The feasible velocity twists (to first order) are of the form

$$k(1, -1, -1, 1, 1, 0), k \in \mathbf{R} \tag{5.40}$$

giving a line in velocity twist space.

5.5 The oriented plane

The previous section developed polyhedral convex cones as a fundamental technique for analyzing a variety of manipulation problems. For spatial problems, the polyhedral convex cones live in the six-dimensional space of wrenches or differential twists. For planar problems, the polyhedral convex cones live in a three-dimensional space of wrenches or differential twists.

But consider Reuleaux's method for analyzing constraint (section 2.5). It represents polyhedral convex cones in differential twist space, and it requires only *two* dimensions, not three. It represents a differential twist by the corresponding rotation center, labeled either ⊕ or ⊖. The technique of labeling points in the plane is the key to Reuleaux's method. The space of labeled planar points is called the *oriented plane*. This section provides a formal definition of the oriented plane, and applies it to represent polyhedral convex cones. Following sections will apply the idea to obtain specific techniques for analyzing planar wrenches and differential twists for planar problems.

We define the oriented plane by using a variation on homogeneous coordinates:

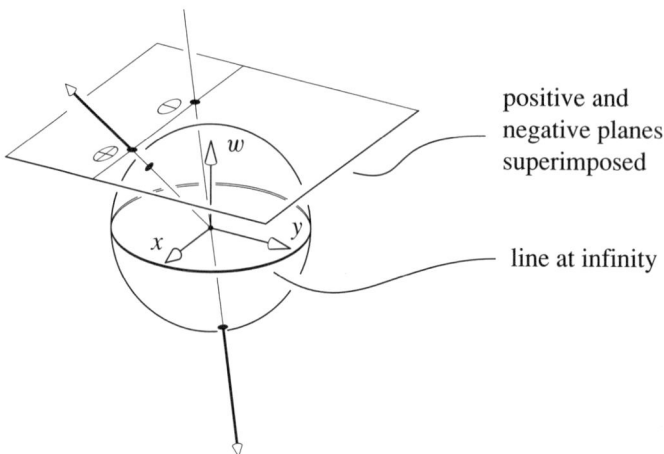

Figure 5.9
The oriented plane.

DEFINITION 5.5: Consider all homogeneous coordinate triples (x, y, w) with x, y, and w all real numbers, and not all simultaneously zero. Each such triple determines a directed line through the origin. A **point in the oriented plane** is a ray of triples:

$$\{(kx, ky, kw) \mid k > 0\} \tag{5.41}$$

all of which give the same directed line. We distinguish three cases:

$w > 0$: For positive w, the ray (kx, ky, kw), $k > 0$, maps to a point in the Euclidean plane with Euclidean coordinates $(x/w, y/w)$ and with a $\oplus$ label. Equivalently, we say it maps to the *positive plane* with coordinates $(x/w, y/w)$.

$w < 0$: For negative w, the ray (kx, ky, kw), $k > 0$, maps to a point in the Euclidean plane with Euclidean coordinates $(x/w, y/w)$ and with a $\ominus$ label. Equivalently, we say it maps to the *negative plane* with coordinates $(x/w, y/w)$.

$w = 0$: For zero w, the ray (kx, ky, kw) is an ideal point. It maps to neither the positive nor the negative plane, but to the ideal line, or line at infinity. For a graphical representation, we can map an ideal point to a point on the unit circle $(x, y)/|(x, y)|$.

The oriented plane is best understood by the central projections illustrated in figure 5.9. (See the appendix for more on central projection and related material on projective geometry.) If we superimpose the positive and negative planes at $w = 1$ in homogeneous coordinate space, then we obtain the correct mapping by intersection. An upward pointing

Rigid Body Statics

($w > 0$) directed line intersects the positive plane at a point labeled $\oplus$. A downward pointing ($w < 0$) directed line intersects the negative plane at a point labeled $\ominus$. A horizontal ($w = 0$) directed line intersects neither plane, and is thus an ideal point, or a point at infinity, represented by intersection with the equator.

Thus we envision the oriented plane as two planes plus a circle. The circle is the glue between the two planes. The planes are connected in such a way that a point moving to infinity in one direction, say the $+x$ axis of the positive plane, reappears from the opposite direction, the $-x$ axis of the negative plane. This is readily observed by using the projection of figure 5.9, and letting some ray in homogeneous coordinate space cross the equator.

Just as the projective plane can be regarded as the set of lines through the origin of $\mathbf{E}^3$, the oriented plane can be regarded as the set of *directed* lines through the origin of $\mathbf{E}^3$. And, just as the projective plane can be regarded as the sphere $S(2)$ with antipodes identified, the oriented plane can be regarded as the sphere $S(2)$ with antipodes *not* identified—just a plain sphere. The northern hemisphere is the positive plane, the southern hemisphere is the negative plane, and the equator is again the ideal line.

GEOMETRY AND CONVEXITY

We can do geometry in the oriented plane. For example, two points determine a line, provided they are not antipodes. We can construct the line by working in the homogeneous coordinate space. Each point of the oriented plane is a ray in homogeneous coordinate space. Unless the points are antipodes, the two rays determine a plane. That plane intersects the sphere in a great circle, and intersects the positive and negative planes in lines. Unless, of course, the two given points are ideal points, in which case we obtain the ideal line.

In practice we work directly with labeled points as Reuleaux did, rather than employing the central projection. We draw both planes superimposed, and label points "$\oplus$" or "$\ominus$" depending on whether they are in the positive or negative plane. When it is necessary to consider ideal points we can use a circle.

Here is the rule for constructing the convex hull of two points in the oriented plane. (See figure 5.10.)

- If the two points are antipodes, that is they have the same coordinates in the plane but opposite signs, the convex hull is just the two points.

- If the two points have the same sign, construct the line segment joining them and give it the same sign as the two points.

- If the two points have opposite sign, construct the line through the two points. The two points divide the line into three parts:

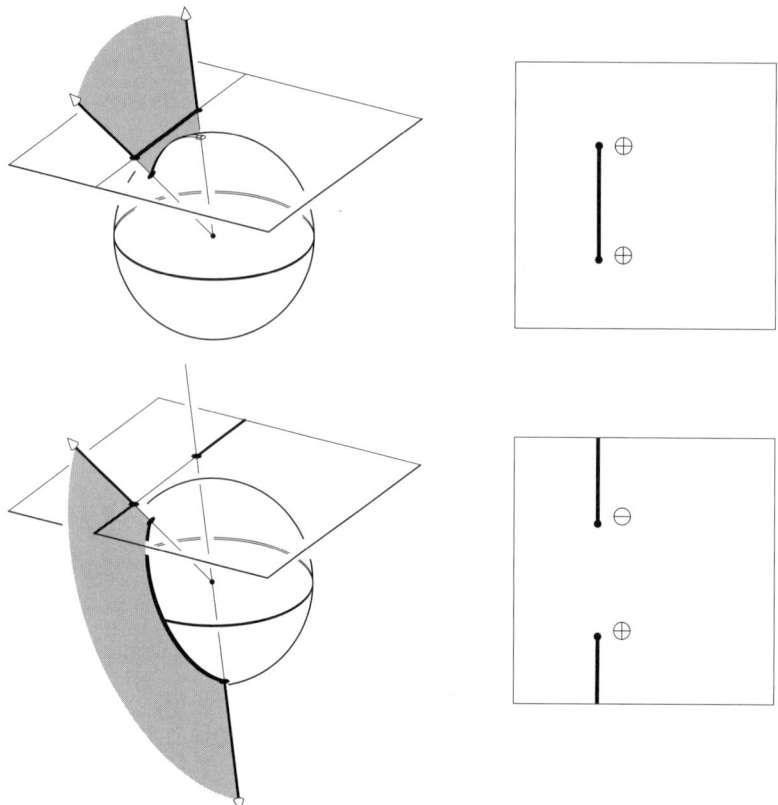

Figure 5.10
Line segments on the oriented plane.

—a ray bounded by a "⊕" point – label that ray "⊕";

—a ray bounded by a "⊖" point – label that ray "⊖";

—a line segment between the "⊕" and "⊖" points – erase that.

There is also a point at infinity connecting the two rays.

There are other cases to deal with, involving points at infinity, but it is actually easier to figure out the rules than it is to remember them, or to write them down.

How do we know these rules are right? By carrying out the corresponding operations on the sphere, and examining the projection onto the oriented plane. Given two unique points in the oriented plane, construct the corresponding rays in homogeneous coordinate space, take the convex hull of the rays, giving a planar cone or a line, then project the results back to the oriented plane. For example, the rule for antipodal points follows easily from observing that the convex hull of anti-parallel rays is just a line, which projects back to the original points.

Using convex hull of points, we can construct convex polygons in the oriented plane. There are some interesting departures from plane geometry. We have already noted that a set of two antipodal points is convex. Another interesting case is the figure with two edges and two vertices (or is it four?) but nonempty interior. See exercise 5.2 for its construction using convex hull in the oriented plane, and compare it with the central projection of figure 5.5g.

Now for the key observation. The mapping from convex polygons in the oriented plane to polyhedral convex cones in $\mathbf{R}^3$ is one to one. The mapping is just the central projection we used in figure 5.9. Thus the oriented plane is a very practical tool for representing polyhedral convex cones in $\mathbf{R}^3$. The rest of this chapter explores variations of this approach. First we will see that Reuleaux's method and the line of force are both instances of this technique. Then we develop new methods: force-dual and moment-labeling.

5.6 Instantaneous centers and Reuleaux's method

Polyhedral convex cones and the oriented plane gives us the two tools we need for graphical solution of a variety of problems. This section shows that Reuleaux's method can be interpreted as using the oriented plane to represent polyhedral convex cones of planar velocity twists. Suppose we have a planar velocity twist (v_x, v_y, ω). The instantaneous velocity center is the point with coordinates $(-v_y/\omega, v_x/\omega)$. The rotation center label, "$\oplus$" or "$\ominus$", is just the sign of ω. With a rotation of the x-y coordinate axes, this transformation is the same central projection we used to get from $\mathbf{R}^3$ to the oriented plane.

Thus two familiar graphical techniques are revealed to be applications of the oriented plane and polyhedral convex cones:

A rotation center represents a ray in planar differential twist space by projecting it to the oriented plane.

Reuleaux's labeled regions represent polyhedral convex cones in planar differential twist space by projecting them to the oriented plane.

We can now interpret each step of Reuleaux's method as follows. (1) For each contact normal we label rotation center half-planes plus and minus. That is, for each contact normal we take the half-space of reciprocal or repelling differential twists. (2) We retain all consistently labeled points. That is, we intersect all the half-spaces to obtain the cone of differential twists.

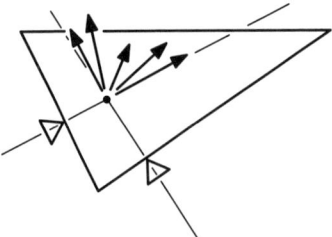

Figure 5.11
Possible resultants of two frictionless contacts.

5.7 Line of force; moment labeling

The previous section showed that two familiar techniques are actually applications of the oriented plane and polyhedral convex cones. This section shows that the line of force is likewise an application of the oriented plane and polyhedral convex cones. This section also introduces a new method called *moment labeling*.

What exactly is a line of force or line of action, first introduced in section 5.1? Given some planar wrench (f_x, f_y, n), we could say that it is the locus of all points at which the moment is zero. But that only gives us a line, not a directed line.

Let's give the line a direction by labeling all the points in the plane either "$\oplus$", "$\ominus$", or "$\pm$", depending on the sign of the moment observed at that point. Every point to the right of the line is labeled "$\ominus$", every point to the left is labeled "$\oplus$", and every point on the line is labeled "$\pm$".

This labeling adds exactly one bit of information to the line. The line bisects the plane. The choice of which half of the plane to label "$\oplus$" and which to label "$\ominus$" determines which direction the line points. This labeling of the plane is just a strange way to draw a directed line: instead of drawing a line with an arrow, draw a line and put a $\oplus$ on the left and a $\ominus$ on the right.

If moment labeling was *only* a strange way to draw a line of force it would not be worth considering. The power of moment labeling becomes evident when we approach more interesting problems. We begin with a problem too simple to require special methods. Suppose we are given a triangle resting on two frictionless support points (figure 5.11). Each support exerts a force normal to the triangle edge. The force can be of arbitrary magnitude, except that it must be directed into the triangle. The question is: what are the possible resultants of the two support forces? In this case there is a simple answer. Slide each support force along its line of action, so that each force acts through the intersection. The resultant force must act through the same point. Now construct the locus of resultant

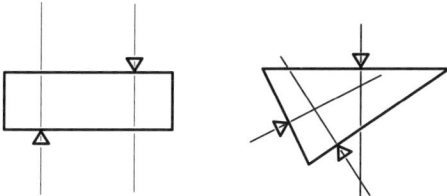

Figure 5.12
Constructing the possible resultants is harder when normals do not have common intersection.

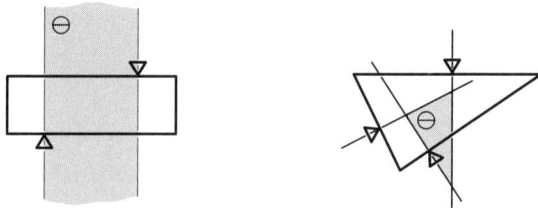

Figure 5.13
Moment labeling adapts Reuleaux's method to represent the possible resultants of planar contacts.

forces, as we vary the magnitudes of the two support forces. The resultant force sweeps out a cone defined by the two lines of action. This cone gives a succinct description of the set of possible resultant forces. Given this construction it is a simple matter to determine the equilibrium position of the triangle under the influence of gravity, and to calculate the contributions of the component support forces.

Now consider two harder problems (figure 5.12). In one case, the two lines of action are parallel. We might be able to apply the simple approach, but it is difficult because the lines of action intersect at infinity. In the other case, there is no hope—the three lines of action do not have a common intersection.

Moment labeling gives a surprisingly simple solution:

1. Use the moment labeling to draw each line of force. (I.e. draw each line of force the strange way—with a $\oplus$ on the left and a $\ominus$ on the right.)
2. Keep all consistently labeled points.

Moment labeling exactly recapitulates Reuleaux's technique, but to represent wrenches instead of differential twists. For each wrench, step 1 gives the half-space of the cone supplementary to that wrench. Step 2 intersects the half-spaces to give a cone which is supplementary to the desired cone of wrenches.

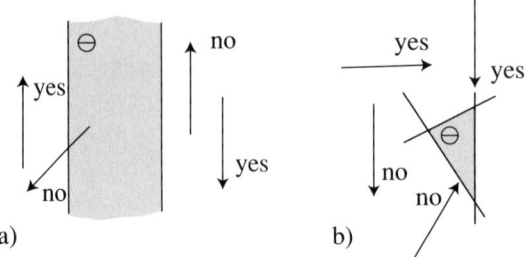

Figure 5.14
How to interpret the moment labeling as a set of possible resultants. In (a) the resultant must be parallel to the shaded strip, directed upward if to the left and downward if to the right. In (b) the resultant must pass the shaded triangle on the left.

How do we interpret such a labeling? There is a simple way to "read" a moment-labeling. A wrench is included in the cone if and only if its line of force passes all ⊕ points on the right, and all ⊖ points on the left (figure 5.14). That is, the line of force has to be between the ⊕ and ⊖ regions, with the right direction. The line of force can touch the boundary of either labeled region, but cannot pass through the interior of either region.

To summarize:

A line of force represents a ray in planar wrench space by projecting its supplementary cone to the oriented plane.

Moment labeling represents a polyhedral convex cone in planar wrench space by projecting its supplementary cone to the oriented plane.

5.8 Force dual

This section introduces one more graphical method for representing cones of wrenches: *force dual*.

Recall that Reuleaux's method represents a cone of twists by central projection to the oriented plane. Moment-labeling represents a cone of wrenches by central projection of the *supplementary* cone to the oriented plane. Why don't we consider the seemingly more straightforward method, and project the cone of wrenches directly to the oriented plane without taking the supplementary cone? That is the *force dual* method.

Force-dual represents a polyhedral convex cone in planar wrench space by central projection to the oriented plane.

Rigid Body Statics

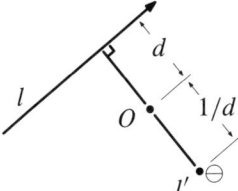

Figure 5.15
Constructing the dual of a line to obtain a point.

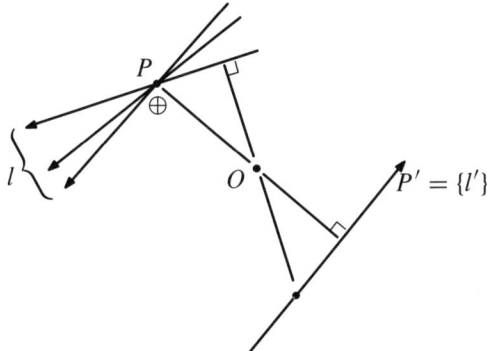

Figure 5.16
Applying the dual transform to a point to obtain a line.

We can develop the force dual method by defining a transformation from a planar wrench to the oriented plane:

$$\begin{pmatrix} f_x \\ f_y \\ n_z \end{pmatrix} \mapsto \begin{pmatrix} -f_y/n_z \\ f_x/n_z \end{pmatrix} \tag{5.42}$$

where the sign of the point is just the sign of the moment n_z. If the moment is zero, we have an ideal point. Thus, as with instantaneous centers, we have a transformation that is just a coordinate rotation away from the projection of figure 5.9.

This transformation has a simple geometrical rendering. Given a force acting along some line, construct the perpendicular to the line, through the origin. The point lies on the perpendicular, on the opposite side of the origin. Its distance from the origin is equal to the inverse distance of the original force to the origin. The third component, the sign, is just the sign of the moment.

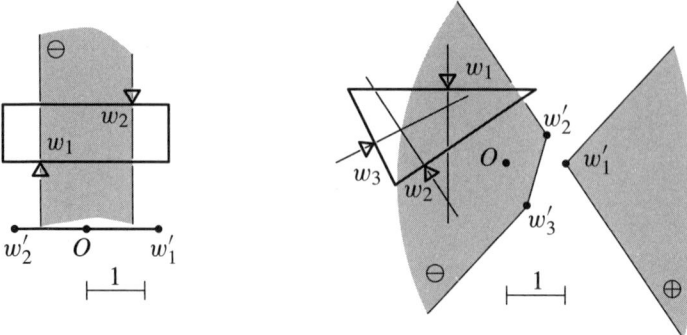

Figure 5.17
Two examples of the force dual method. Two contacts map to a line segment in the oriented plane; three contacts map to a triangle in the oriented plane.

The mapping, as defined, is from directed lines to signed points—the image of a particular wrench is independent of the magnitude, depending only on the (directed) line of force. But we can extend the mapping, so that it also maps signed points back to directed lines. We will define the image of a point P to be the locus of images of all the lines through P, (figure 5.16). Suppose $\{l\}$ is the set of directed lines through P. Then P' is defined to be $\{l'\}$, with a direction determined by the sign of P. A simple geometric construction suffices to show:

- P' is a directed line
- $P'' = P$

Hence the transformation is *dual*, which is how the the name *force dual* arose.

For example, let's apply the force dual method to the problems of figure 5.12. To apply the force dual method:

1. Choose the origin and unit length wisely.
2. For each directed line of action w_i construct its dual in the oriented plane w'_i.
3. Take the convex hull conv($\{w'_i\}$).

The result is a convex figure in the oriented plane, which represents the positive linear span of the given w_i. Each point in the figure is the dual of a possible resultant.

A "wise" choice of the origin and unit length means that you should anticipate where the dual constructs will go, and keep them on the page. Graphical methods are awkward when all the action is at infinity. So don't put the origin right on top of the contact normals. At the same time, it is convenient if the dual constructs are not right on top of the original

Rigid Body Statics

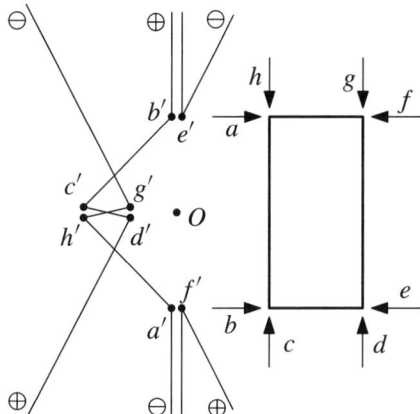

Figure 5.18
The force dual transform maps every contact normal to a point in the oriented plane. The resulting curve is called the "zigzag locus".

figure. The placement of origins in figure 5.17 may seem counterintuitive at first, but note that the main features of the dual figures are on the page and not on top of the original figures.

The first example of figure 5.17 has two anti-parallel contact normals. By placing the origin between them, the moments of the normals have the same sign ⊖. The convex hull is a simple line segment also labeled ⊖. Compare this line segment with the shaded region in figure 5.13 (moment labeling) or 2.19(a) (Reuleaux's method). Each vertex in one figure is dual to an edge in the other. The figures are projections of supplementary cones, but are different ways of representing the same set.

The second example shows three contact normals of a triangle, mapping to three points in the oriented plane. Their convex hull gives a triangle in the oriented plane, which overlaps the line at infinity. Again comparison with moment labeling (figure 5.13) and Reuleaux's method (figure 2.19(b)) is instructive.

Since the force dual method is a little more involved than the moment-labeling method, you might wonder why we need it. The beauty of the dual mapping is that it represents each (directed) line of action as a (signed) point. So that a *set* of lines of action is represented by a *region*. Of course, the moment labeling method also represents a set by a region, but not in the usual sense that a region refers to a set of points. Because of this, the force dual method can represent an arbitrary set of lines of action, not just convex cones. For example, suppose we want to represent the set of frictionless forces that could be applied to the perimeter of an object (figure 5.18). This is easily described by a piecewise-linear

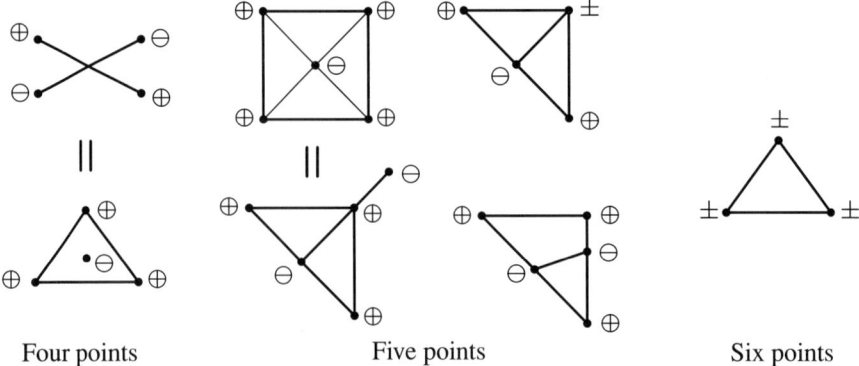

Figure 5.19
Irreducible arrangements of oriented points yielding closure.

closed curve in force dual space, called the *zigzag locus*. But it is not convex, because we did not ask for the possible *resultants* of all such forces, and therefore did not take the convex hull in the oriented plane. As an exercise, the reader might consider the problems of representing this by moment-labeling.

The force dual method is ideal for problems in force closure and first-order form closure. We have already seen some examples, and there are more examples in the next section and in the exercises. The force dual method is also useful when we include friction, as we shall see in the next chapter.

The force dual method will also be useful for dynamic problems. In fact, as we shall see in chapter 8, the force dual transformation arises naturally from Newton's laws.

One thing that the force dual method is *not* good for is visualization. It is not immediately evident what wrenches are included in a polyhedral convex cone represented using force dual. When necessary the dual transform can be used to obtain the equivalent moment labeling representation. When a computer is doing the work, Brost (1991a) found that central projection onto the sphere is the most effective way of visualizing polyhedral convex cones in planar wrench space or differential twist space.

Arrangements yielding closure in the oriented plane

Force closure and first order form closure are represented in the oriented plane by a set of points whose convex hull exhausts the entire oriented plane. We have seen some examples of closure, but we might ask the more general question: what are the different arrangements of points in the oriented plane, such that the convex hull is the entire oriented plane? Can we systematically identify every topologically distinct arrangement that cannot be reduced

Rigid Body Statics

to a smaller set? Figure 5.19 shows all the irreducible arrangements I could identify. Any arrangement of points yielding closure should contain one of the arrangements shown in the figure.

5.9 Summary

The graphical techniques presented above can be summarized in a table:

	Project cone to oriented plane	**Project supplementary cone to oriented plane**
single wrench	(acc'n center)	line of action
single diff'l twist	IC	?
wrench cone	force dual	moment labeling
diff'l twist cone	Reuleaux	?

The table notes that the force dual method, applied to a single wrench, produces the *acceleration center* which is described in section 8.9. Also, we see room in the table for one more method labeled "?", which might be called "velocity dual" or "differential twist dual".

5.10 Bibliographic notes

The material on systems of forces acting on rigid bodies, and the related equivalence theorems, was adapted from (Symon, 1971). Fundamental results on wrenches came from (Ohwovoriole, 1980), (Roth, 1984), and (Hunt, 1978). Polyhedral convex cones are developed in(Goldman and Tucker, 1956). The oriented plane is described by Stolfi (1988). Also see the earlier paper with Guibas and Ramshaw (1983).

Modelling kinematic constraint and systems of contact forces by linear inequalities in wrench space has developed through a number of papers addressing grasp planning, work-holding fixture design, and robotic assembly (Erdmann, 1984; Asada and By, 1985; Kerr and Roth, 1986; Rajan et al., 1987; Mishra et al., 1987; Brost, 1991b; Hirai and Asada, 1993). Erdmann (1984) employed cones in wrench space, and Nguyen (1988) and Hirai and Asada (1993) extended and further developed the use of cones. The force dual and moment labeling methods were described by Brost and myself (1989; 1991). Readers wishing to push deeper into the kinematics and statics of contact should also look into the formulation as a linear complementarity problem. (Pang and Trinkle, 1996) is a good place to start.

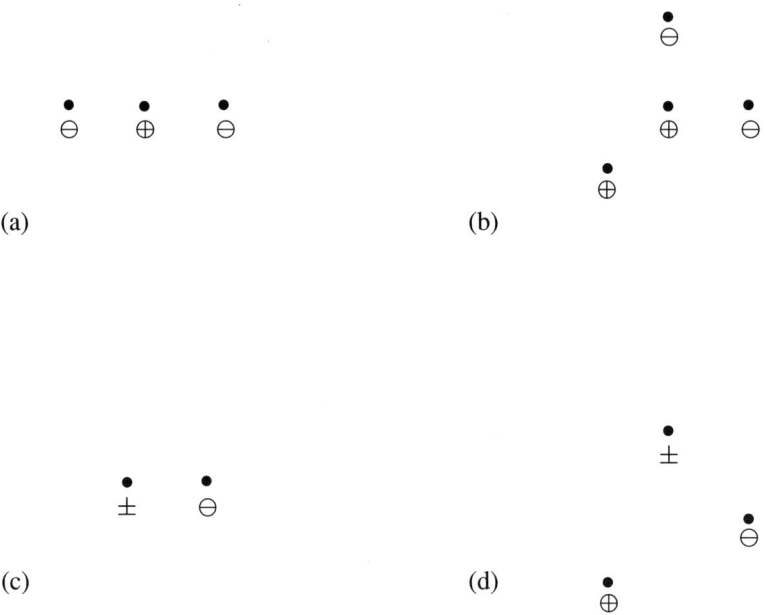

Figure 5.20
Convex hull problems for exercise 5.2.

Exercises

Exercise 5.1: Figure 5.6 shows all the different types of cones in $\mathbf{E}^2$, and the corresponding supplementary cones. Construct an equivalent figure for $\mathbf{E}^3$, showing the supplementary cone for each cone of figure 5.5.

Exercise 5.2: For each set of points in the oriented plane in figure 5.20, construct the convex hull.

Exercise 5.3: Figure 5.21 shows six objects with frictionless contacts. For each, use moment labeling to determine the set of possible resultant wrenches. In each case show the line of action of one of the possible resultant wrenches.

Exercise 5.4: Repeat exercise 5.3, using the force dual method, rather than moment labeling. For each of the six objects, find the force dual representation of the possible resultant wrenches. In each case, choose one of the labeled points in the force dual representation, and apply the dual transform to get the corresponding directed line of action.

Rigid Body Statics

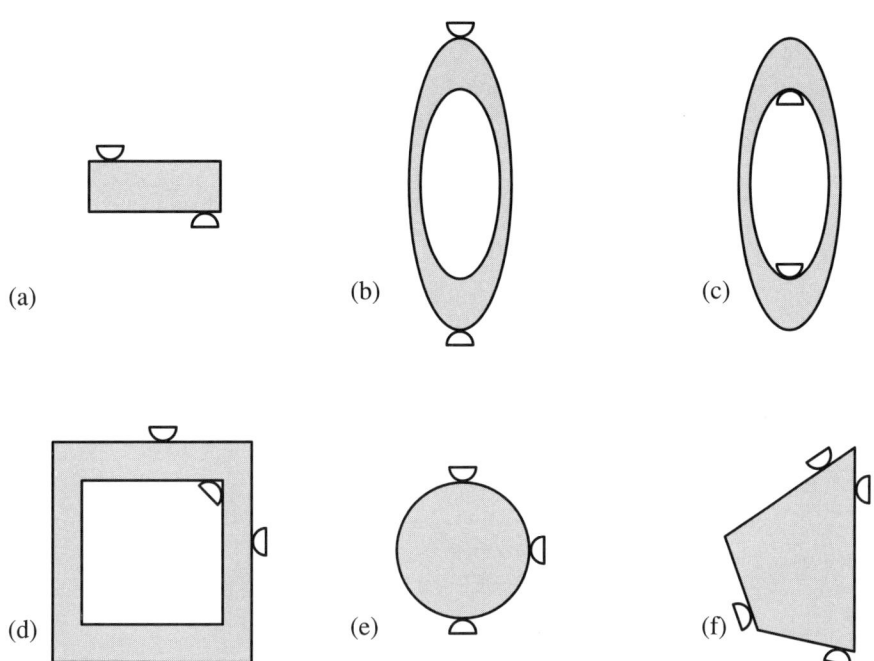

Figure 5.21
Contact problems for exercises 5.3 and 5.4.

Exercise 5.5: For each of the moment-labelings of figure 5.22, four wrenches are drawn. Indicate for each wrench whether or not it is in the wrench cone.

Exercise 5.6: Show that the dual mapping defined by equation 5.42 really does yield the point on the perpendicular to the line of force through the origin, at a distance equal to the inverse of the moment arm. (Recall that the line of force can be defined as the locus of points **x** at which the moment is zero.)

Exercise 5.7: For the triangle of figure 5.13, use the force dual method to determine all placements for a fourth finger to obtain force closure. You should include both types of fingers: pointy fingers on triangle edges, and flat fingers on triangle vertices.
(a) Construct the triangle of force duals for the three given fingers.
(b) Construct the locus of force duals for all contact normals (the zigzag locus).
(c) Identify those duals which, added to the three given contacts, will produce a convex hull that exhausts the oriented plane.

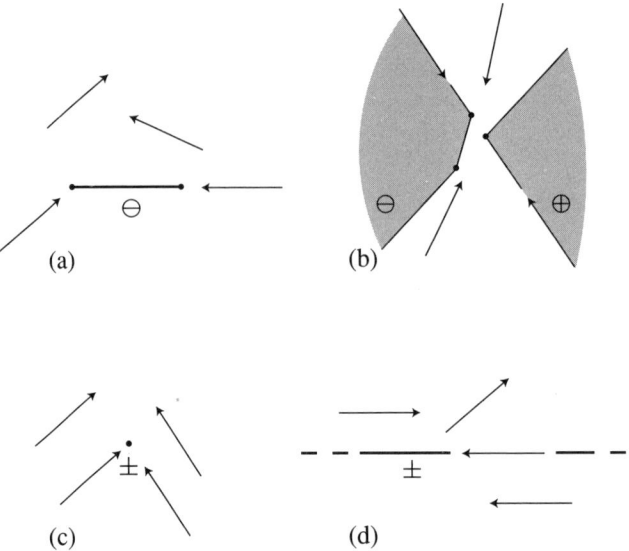

Figure 5.22
Moment labeling interpretation problems for exercise 5.5.

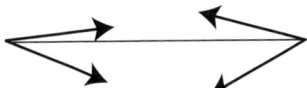

Figure 5.23
Facing cones (exercise 5.8).

(d) Transform back to obtain a subset of the contact normals.

(e) Describe the corresponding set of finger placements.

Exercise 5.8: Nguyen (1988) observes that four wrenches can provide closure if and only if they can be arranged into two pairs, defining two cones each of which contains the other's base. A set of wrenches satisfying Nguyen's criterion is shown in figure 5.23. Show that by suitable placement of the origin, the force dual representation of the wrenches in figure 5.23 will match both of the four point arrangements shown in figure 5.19.

Exercise 5.9: For each of the arrangements of figure 5.19, construct a shape and a set of contact normals with force duals matching the arrangement.

6 Friction

Consider an ordinary manipulation task—a mundane task such as we face every day. The problem of shuffling paper at one's desk, cooking a meal, or playing a game of cards. For any of these tasks, some reflection will yield the following observations:

- most of the objects are at rest most of the time;
- motion of an object generally requires that the hand apply some force.

In short, the mechanics of manipulation is closer to the Aristotelian model than to the Newtonian model. Objects in motion actually do tend to come to rest, in the world of friction and plastic impact. And it is quite fortunate that this is so. Manipulation is often the process of using just a few motors to adjust the positions of many different objects. This is accomplished by moving the objects one, or a few, at a time. It is crucial that the other objects stay put.

Also, friction is often the means by which the hand applies forces to an object. For example, to grasp an object securely without friction requires that the object be surrounded on all sides. With friction, two fingers on opposing features will often do the trick. (But for most purposes two fingers is the lower bound; hence the Hopi aphorism "One finger cannot lift a pebble.")

6.1 Coulomb's Law

We will begin the study of friction by reviewing Coulomb's law. Consider a simple experiment, involving a block sliding on a horizontal surface, pulled by a string. We will assume that the two surfaces are fairly clean, dry, and unlubricated. Imagine that we have instruments to measure the force $\mathbf{f}_a$ applied via the string, and the tangential force $\mathbf{f}_t$ due to friction between the table and the block. If we gradually increase the applied force $\mathbf{f}_a$, we tend to see the behavior illustrated in figure 6.1. For small applied forces, the frictional force will balance the applied force, so that the block does not move. Above some threshold, the block will begin moving, and the frictional force will now be constant. If we conduct a large number of experiments, varying the block's mass, the block's shape, the materials, and so forth, we will find that the limiting frictional forces depend on the normal force, and on the materials involved, but are virtually independent of other factors. If f_{ts} is the limiting value of static friction, at which motion begins, and f_{td} is the value of

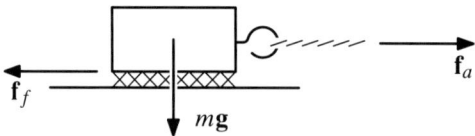

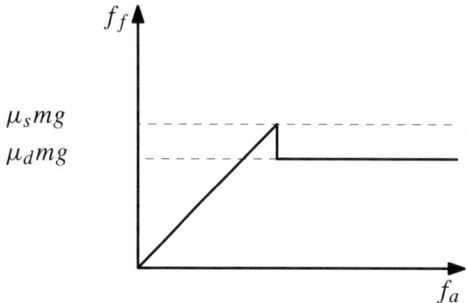

Figure 6.1
Coulomb's law of sliding friction.

dynamic friction, then these values are approximately:

$$f_{ts} = \mu_s f_n \tag{6.1}$$
$$f_{td} = \mu_d f_n \tag{6.2}$$

where f_n is the normal force, μ_s is the *static coefficient of friction*, and μ_d is the dynamic coefficient of friction. Typically μ_s is larger than μ_d, but we will ignore this difference, and speak of a single coefficient of friction μ. A simple statement of Coulomb's law is:

$$|\mathbf{f}_t| \leq \mu |\mathbf{f}_n| \tag{6.3}$$

if there is no motion, and

$$|\mathbf{f}_t| = \mu |\mathbf{f}_n| \tag{6.4}$$

if there is motion, with the tangential force in a direction opposite to the motion.

In particular, the tangential force is largely independent of the contact area, and of the velocity of motion. The coefficient of friction is considered to be a material property, depending only on the materials involved. Tables of the coefficient of friction should not be taken too seriously, but some typical values are given below:

Friction

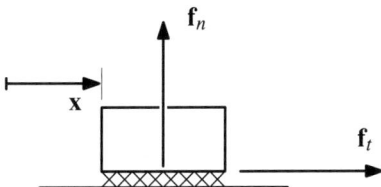

Figure 6.2
Sliding block with Coulomb friction.

Materials	μ
metal on metal	0.15–0.6
rubber on concrete	0.6–0.9
plastic wrap on lettuce	∞
Leonardo's number	0.25

Experiments like those described above provide the basis of Coulomb's law, which will be stated more carefully later. The history of the law is also interesting. Coulomb's work with friction was his first scientific achievement. Coulomb's interest in friction was spurred by practical engineering matters. He was a career military engineer, and carried his huge laboratory apparatus with him from one assignment to another. Coulomb is also known for the invention of the torsion balance and for his studies of electricity, leading to the law for electrostatic attraction which is unfortunately also known as Coulomb's law.

Coulomb was not the first to propose Coulomb's law of sliding friction. Amontons had proposed it earlier, and the law is occasionally referred to as Amontons's law. It also appears that Leonardo da Vinci had earlier posed a more restrictive version of the law, supposing the coefficient of friction to be always one fourth. Coulomb's law is a phenomenological law, providing an approximate description of an aggregate behavior. For that reason there are some who would prefer not to call it a law at all. There are more fundamental approaches to the modeling of friction, and more precise approximations of friction. But Coulomb's law still provides the best combination of simplicity and accuracy for many purposes.

6.2 Single degree-of-freedom problems

We begin by considering the simplest problems, involving just one degree of freedom. Consider a block in frictional contact with a supporting plane, which is prevented somehow from moving away from the plane (figure 6.2). The tangential position of the block is given by x, and the frictional force is given by f_n and f_t, respectively normal and tangential to the support plane. Coulomb's law prescribes a constraint on the contact force, depending on the *contact mode* indicated in the table below:

$\dot{x}$	$\ddot{x}$		
< 0		$f_t = \mu f_n$	left sliding
> 0		$f_t = -\mu f_n$	right sliding
$= 0$	< 0	$f_t = \mu f_n$	left sliding
$= 0$	> 0	$f_t = -\mu f_n$	right sliding
$= 0$	$= 0$	$|f_t| \leq |\mu f_n|$	rest

Now suppose that we introduce a gravitational field so that the supporting surface is an inclined plane (figure 6.3). Let α be the angle of the inclined plane with respect to the horizontal. What is the maximum angle α at which the block can remain at rest?

If the block is at rest, then the gravitational force must balance the total contact force:

$$f_n = mg \cos \alpha \tag{6.5}$$

$$f_t = mg \sin \alpha \tag{6.6}$$

At rest we have $|f_t| \leq |\mu f_n|$. The limiting case is given by

$$f_t = \mu f_n \tag{6.7}$$

Substituting,

$$mg \sin \alpha = \mu mg \cos \alpha \tag{6.8}$$

$$\alpha = \tan^{-1} \mu \tag{6.9}$$

The desired angle α is the arc-tangent of the coefficient of friction. This angle is sometimes called the *friction angle* or the *angle of repose*.

The friction angle provides an elegant geometrical approach to Coulomb's law. Consider all the forces satisfying Coulomb's law for an object at rest, i.e. all the forces satisfying the condition

$$|f_t| \leq \mu |f_n| \tag{6.10}$$

This set of forces describes a cone in the force space, called the *friction cone*, with vertex at the origin, and dihedral angle $2 \tan^{-1} \mu$ (figure 6.4). Then we can state Coulomb's law:

> For left sliding $\mathbf{f}_n + \mathbf{f}_t \in$ right edge of friction cone
> For right sliding $\mathbf{f}_n + \mathbf{f}_t \in$ left edge of friction cone
> For rest $\mathbf{f}_n + \mathbf{f}_t \in$ friction cone

Friction

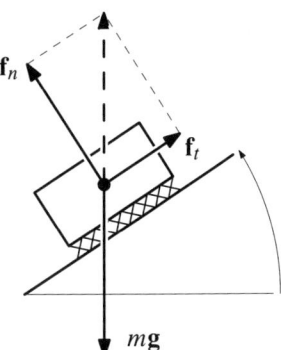

Figure 6.3
Sliding block on inclined plane.

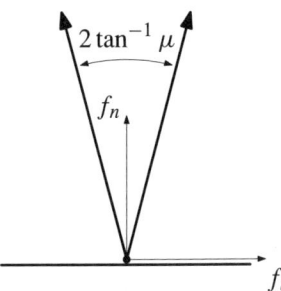

Figure 6.4
The friction cone: to satisfy Coulomb's law the contact force deviates by at most $\tan^{-1}\mu$ from the contact normal.

To illustrate the friction cone, consider a problem in pipe clamp design (figure 6.5). The peculiar property of a pipe clamp is that, although there is complete freedom of movement, a force applied at the face of the clamp will not generate a motion. The clamp jams, and increasing the force just increases the balancing force.

Now, consider the problem of designing such a clamp. In particular, where do we place the face of the clamp? How far must the face be from the pipe center, to obtain the jamming effect? For the purposes of the example, let the diameter of the pipe be 2 cm, let the length of the sliding element be 2 cm, and let the coefficient of friction be 0.25.

The relevant forces are shown in the figure. The applied force acts through the face, and is of arbitrary magnitude. The relevant contact forces are described by two friction cones.

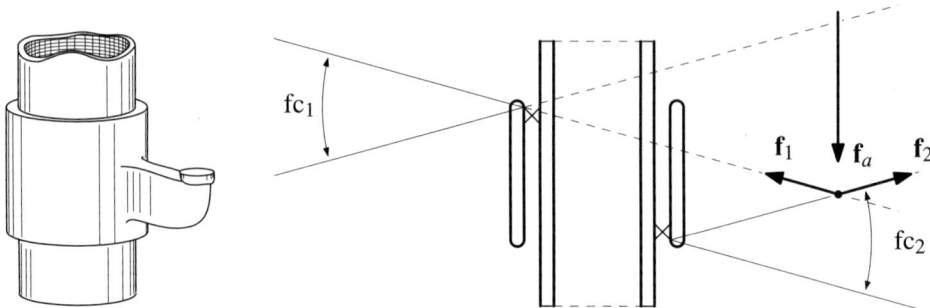

Figure 6.5
Analysis of the pipe clamp: a force applied far enough from the pipe causes the clamp to jam.

These two forces are also of arbitrary magnitude, but they must lie within their respective friction cones.

Now for the clamp to remain at rest, the forces must be in equilibrium. A necessary condition is that all three forces pass through a common point. Since the two contact forces must lie within their friction cones, then the applied force must pass through the *intersection* of the two friction cones. This condition is easily satisfied by placing the face in the intersection of the two friction cones. The relevant constructions are shown in the figure. The minimum distance is 4 cm from the pipe center. We also note that with the face in the intersection of the two friction cones, i.e. about halfway along the length of the sliding element, the effect can tolerate large variations in the direction of the applied force.

6.3 Planar single contact problems

The previous section assumes that a sliding object will remain in contact with the supporting surface, yielding just three contact modes: left sliding, right sliding, and rest. More generally, an object might move away, losing the contact. As a result we have some additional contact modes. Consider a simple rod in contact with a fixed surface. Let $\{\hat{\mathbf{t}}, \hat{\mathbf{n}}\}$ be a coordinate system placed at the contact point $\mathbf{p}_c$, with $\hat{\mathbf{t}}$ tangential to the contact, and $\hat{\mathbf{n}}$ normal to the contact.

Friction

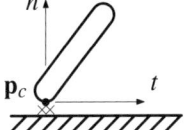

Figure 6.6
Sliding rod with Coulomb friction.

Then the possible contact modes are enumerated below.

$\dot{p}_{cn}$	$\ddot{p}_{cn}$	$\dot{p}_{ct}$	$\ddot{p}_{ct}$		
< 0					impact
> 0					separation
= 0	< 0				impact
= 0	> 0				separation
= 0	= 0	< 0		$f_t = \mu f_n$	left sliding
= 0	= 0	> 0		$f_t = -\mu f_n$	right sliding
= 0	= 0	= 0	< 0	$f_t = \mu f_n$	left sliding
= 0	= 0	= 0	> 0	$f_t = -\mu f_n$	right sliding
= 0	= 0	= 0	= 0	$|f_t| \leq |\mu f_n|$	fixed

The first four lines identify cases where the bodies are approaching each other, resulting in an impact (chapter 9), or where the bodies are separating, giving zero contact forces. The remaining five lines essentially repeat the table of the previous section. There is one difference, though. The earlier table had a mode labeled "rest" which is now labeled "fixed". The rod need not be at rest—it might be rotating about the contact point. More generally, if the end of the rod were round, a rolling contact could give "fixed" contact, even though the contact point is in motion. For this reason, this contact mode is sometimes referred to as "fixed or rolling" in the literature. We will not consider rolling contacts further.

6.4 Graphical representation of friction cones

We can immediately apply either the moment labeling method or the force dual method to friction problems (figure 6.7). The moment labeling of a friction cone is just a slight variation in the usual method of drawing a friction cone—drawing the outside instead of the inside of the cone. The force dual method produces a line segment in force dual space.

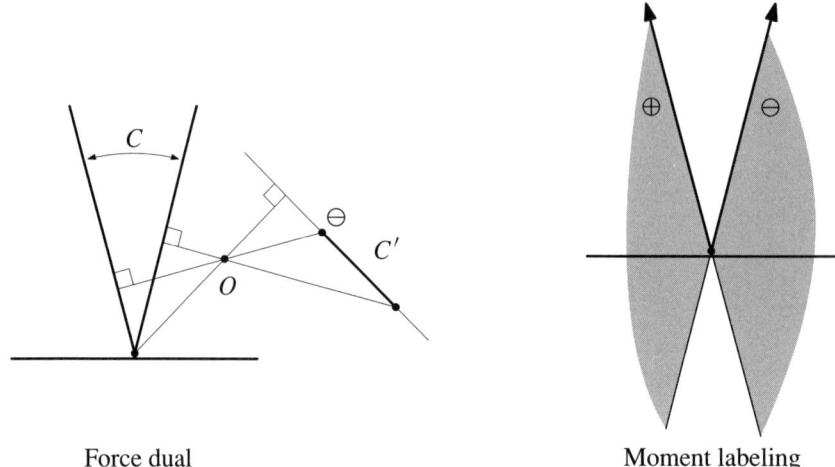

Force dual   Moment labeling

Figure 6.7
Representing the friction cone by force dual or moment labeling.

6.5 Static equilibrium problems

Static equilibrium is a simple condition to analyze, but it is very useful. For example, many problems in grasp synthesis, in complex mechanical assemblies, in work-holding fixtures, can be addressed by simple static equilibrium.

Assuming a rigid object initially at rest, we define static equilibrium to mean that the static forces—contact forces and applied load forces—sum to zero. Recall that to define force, we assumed an object at rest with zero total force would remain at rest. However, it is also advisable to consider the effect of perturbations to the object, augmenting static equilibrium analysis with *stability* analysis.

EXAMPLE 1: BLOCK ON TABLE

Consider a block at rest on a horizontal surface (figure 6.8). The forces acting on the block are the contact forces from the table, and a gravitational load. We can check for static equilibrium by constructing the set of possible resultants of contact wrenches $\text{pos}(\{c_i\})$, and then see if one of these resultants will balance the gravitational load wrench w_g. In short, static equilibrium holds if $-w_g \in \text{pos}(\{c_i\})$. The relevant constructions are given in the figure, for moment labeling and force dual representations. Notice that the set of contact wrenches is a continuum, but is equivalent to just two contacts, at the ends of the contact edge. The block is stable—if we perturb the block it quickly comes to rest at a nearby position.

Friction

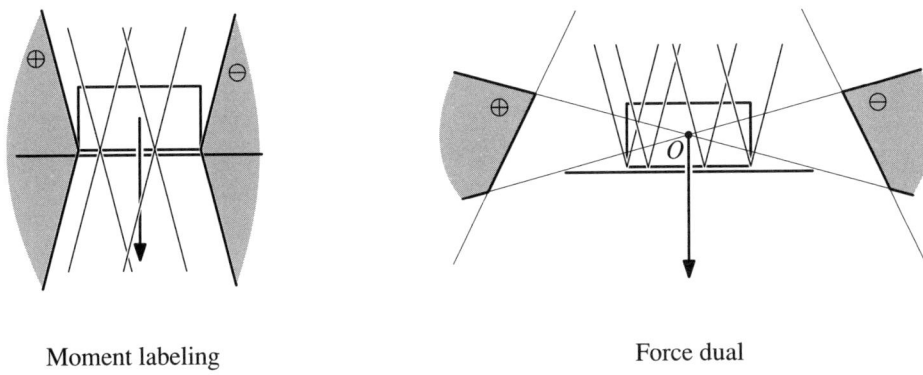

Moment labeling   Force dual

Figure 6.8
Moment labeling represents friction cone for block on table.

Figure 6.9
Will it fall? With nonzero friction this beam is in force closure, but it is not necessarily in equilibrium.

EXAMPLE 2: FORCE CLOSURE AND INDETERMINACY

It seems obvious that force closure implies static equilibrium, but that is not the case. Force closure implies that the possible contact wrench resultants exhaust all of wrench space. Thus they *can* balance any applied force, but *will* they? In figure 6.9 the coefficient of friction is not zero, and the beam caught in the chasm is in force closure. Although there is a choice of contact forces that will balance the gravitational load, there are other choices of contact forces that also satisfy the laws of rigid body mechanics. In particular, if the contact forces are zero, the beam will fall. If we want to predict whether the beam will fall or not, we have to go beyond rigid body mechanics, and model the beam or walls as

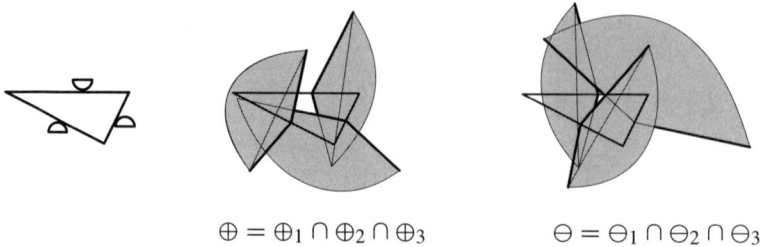

$$\oplus = \oplus_1 \cap \oplus_2 \cap \oplus_3 \qquad \ominus = \ominus_1 \cap \ominus_2 \cap \ominus_3$$

Figure 6.10
The three positive regions have no common intersection, nor do the three negative regions. Hence the labeled regions are null, the possible resultants are all of wrench space, and the triangle is in force closure.

deformable bodies. For the problem as stated, we have force closure, we *may* have static equilibrium, but we surely would not classify it as stable.

EXAMPLE 3: A TRIANGLE IN FORCE CLOSURE

Consider a three-finger grasp of a triangle in the plane (figure 6.10). Kinematic analysis, or force analysis using frictionless contact, suggest that the triangle is not securely grasped, that is, that there are some loads that cannot be balanced by the contact forces. However, with sufficient frictional forces, the possible resultants exhaust all of wrench space, and the triangle is in force closure. The figure shows the moment labeling of the problem. The reader may easily verify that both the positive-labeled and the negative-labeled regions are empty. Thus there are no constraints on the possible wrenches: force closure.

6.6 Planar sliding

Some manipulation tasks involve an object sliding on a planar support surface. The mechanics of planar sliding apply to problems as diverse as moving furniture (section 7.2) and fixturing objects for machining operations. This section develops expressions for the frictional force and moment of planar sliding, and introduces an elegant graphical representation known as the *Limit Surface*.

The motion of a pushed object is often indeterminate. If a rigid object is supported by more than three contact points, the distribution of support forces is underdetermined. If the frictional forces are assumed proportional to the normal forces, as Coulomb suggests, then the frictional forces are also underdetermined. The problem is illustrated by the defective dinner plate of figure 6.11. The plate was designed with a circular ridge on the bottom, so that the support forces would be concentrated at the edge of the plate. Unfortunately, the bottom of the plate sagged during the firing process, so that the center is also in contact

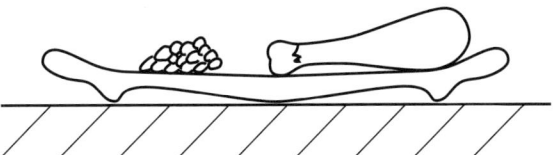

Figure 6.11
It is impossible to predict the motion of this plate, without knowing the distribution of support forces between the plate and the table.

with the planar support. There is no way to predict whether the support forces will be concentrated at the center, giving it an irritating tendency to rotate, or at the edge, resisting rotation. In practice, the plate's behavior will depend on details that may be very difficult to model. It may behave well with a tablecloth, and poorly without a tablecloth. Its behavior might depend on the phase of the moon. (Tidal forces induce microscopic changes in the shapes of the plate and table.)

The defective dinner plate is a particularly egregious example of the indeterminacy of planar sliding. In the worst case the problem can be very awkward, but in most practical situations there are many ways of addressing the problem. One inescapable conclusion is that a useful theory of planar sliding should capture this indeterminacy, which is a primary goal for the approach described below.

The first step is to develop expressions for the force and moment of planar sliding under the assumption that the support forces are known and described by a finite *pressure distribution* $p(\mathbf{r})$. Under those assumptions indeterminacy is not an issue—there is a one-to-one mapping between the direction of slider motion and the resulting wrench, except when the slider is motionless.

Given the force and moment for a known finite pressure distribution, the next step is to generalize to cases where there may be finite force concentrated at an isolated point of support, corresponding to infinite pressure. In those cases the mapping between direction of slider motion and the resulting wrench may be many-to-one or one-to-many.

Finally, we also have to consider the indeterminacy arising from an unknown or partially known pressure distribution, such as the dinner plate example above, which is addressed in chapter 7.

Force and moment of planar sliding

Let some object be in planar motion, supported by a fixed planar surface. Choose a coordinate frame with the x-y plane coincident with the support, and z pointing outward. Let the object's contact with the surface be confined to some region R. Let $\mathbf{r}$ be the position vector of some point in the object, and let $\mathbf{v}(\mathbf{r})$ be the velocity of that point. If $p(\mathbf{r})$ is the

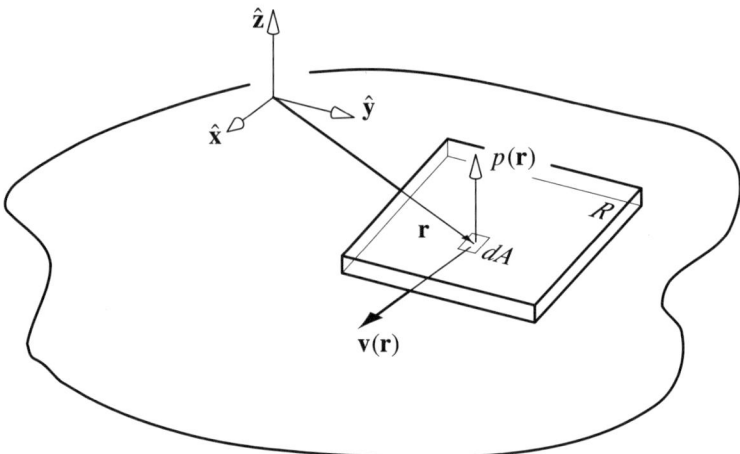

Figure 6.12
Notation for planar sliding.

pressure at **r**, and dA a differential element of area at **r**, then the magnitude of the normal force at **r** is given by

$$p(\mathbf{r})\, dA \tag{6.11}$$

and Coulomb's law gives us the tangential force at **r**:

$$-\mu \frac{\mathbf{v}(\mathbf{r})}{|\mathbf{v}(\mathbf{r})|} p(\mathbf{r})\, dA \tag{6.12}$$

for $|\mathbf{v}(\mathbf{r})| \neq 0$, where μ is the coefficient of friction, assumed uniform over the contact region R.

Integrating over R, we obtain expressions for the total force and moment due to friction:

$$\mathbf{f}_f = -\mu \int_R \frac{\mathbf{v}(\mathbf{r})}{|\mathbf{v}(\mathbf{r})|} p(\mathbf{r})\, dA \tag{6.13}$$

$$\mathbf{n}_f = -\mu \int_R \mathbf{r} \times \frac{\mathbf{v}(\mathbf{r})}{|\mathbf{v}(\mathbf{r})|} p(\mathbf{r})\, dA \tag{6.14}$$

Note that the frictional force $\mathbf{f}_f$ lies in the x-y plane, and the total frictional moment $\mathbf{n}_f$ acts along the z-axis. Without knowledge of the pressure distribution $p(\mathbf{r})$, these integrals cannot be evaluated, leading to indeterminacy in the frictional forces. There is an exception, though: pure translation.

CASE 1: PURE TRANSLATION

If the object is in pure translation, all points are moving in the same direction, and we can factor the integrals of equations 6.13 and 6.14.

$$\mathbf{f}_f = -\mu \frac{\mathbf{v(r)}}{|\mathbf{v(r)}|} \int_R p(\mathbf{r})\, dA \tag{6.15}$$

$$\mathbf{n}_f = -\mu \int_R \mathbf{r}\, p(\mathbf{r})\, dA \times \frac{\mathbf{v(r)}}{|\mathbf{v(r)}|} \tag{6.16}$$

Let $\mathbf{f}_0$ be the total normal force, and let $\mathbf{r}_0$ be the centroid of the pressure distribution. Then

$$f_0 = \int_R p(\mathbf{r})\, dA \tag{6.17}$$

$$\mathbf{r}_0 = \frac{1}{f_0} \int_R \mathbf{r}\, p(\mathbf{r})\, dA \tag{6.18}$$

Substituting into equations 6.15 and 6.16,

$$\mathbf{f}_f = -\mu \frac{\mathbf{v(r)}}{|\mathbf{v(r)}|} f_0 \tag{6.19}$$

$$\mathbf{n}_f = \mathbf{r}_0 \times \mathbf{f}_f \tag{6.20}$$

Hence the frictional forces distributed over the support region have a resultant, with magnitude μf_0, in a direction opposing the motion, through the centroid $\mathbf{r}_0$. In other words, the force is equivalent to that obtained by applying Coulomb's law to the sliding of a single point located at $\mathbf{r}_0$.

DEFINITION 6.1: The **center of friction** is the centroid $\mathbf{r}_0$ of the pressure distribution.

THEOREM 6.1: For a rigid body in purely translational sliding on a planar surface, with uniform coefficient of friction, the frictional forces reduce to a force through the center of friction, opposing the velocity.

Proof Given above. ∎

In some cases, the center of friction is easily determined. If an object is at rest on the support plane, with no applied forces other than gravity and the support contact forces, then the center of friction is directly below the center of gravity. This is the only location that allows the contact forces to balance the gravitational force. We can generalize slightly, allowing additional applied forces, as long as they are in the support plane. We can also permit accelerated motion of the body, if the center of gravity is in the support plane. But acceleration of a body whose center of gravity is above the support plane will, in general, cause a shift in the pressure distribution, and a corresponding shift in the center of friction. Applied forces not lying in the support plane will generally cause a similar shift.

CASE 2: ROTATION

Now suppose that the body is rotating, with an instantaneous center $\mathbf{r}_{IC}$. Then the velocity of a point at $\mathbf{r}$ is given by

$$\mathbf{v}(\mathbf{r}) = \boldsymbol{\omega} \times (\mathbf{r} - \mathbf{r}_{IC}) \tag{6.21}$$

$$= \dot{\theta}\,\hat{\mathbf{k}} \times (\mathbf{r} - \mathbf{r}_{IC}) \tag{6.22}$$

and the direction of motion at $\mathbf{r}$ is

$$\frac{\mathbf{v}(\mathbf{r})}{|\mathbf{v}(\mathbf{r})|} = \operatorname{sgn}(\dot{\theta})\,\hat{\mathbf{k}} \times \frac{\mathbf{r} - \mathbf{r}_{IC}}{|\mathbf{r} - \mathbf{r}_{IC}|} \tag{6.23}$$

Substituting into equations 6.13 and 6.14 we obtain

$$\mathbf{f}_f = -\mu\,\operatorname{sgn}(\dot{\theta})\,\hat{\mathbf{k}} \times \int_R \frac{\mathbf{r} - \mathbf{r}_{IC}}{|\mathbf{r} - \mathbf{r}_{IC}|} p(\mathbf{r})\,dA \tag{6.24}$$

$$n_{fz} = -\mu\,\operatorname{sgn}(\dot{\theta}) \int_R \mathbf{r} \cdot \frac{\mathbf{r} - \mathbf{r}_{IC}}{|\mathbf{r} - \mathbf{r}_{IC}|} p(\mathbf{r})\,dA \tag{6.25}$$

Notice that these equations have a well-defined limit as the rotation center $\mathbf{r}_{IC}$ approaches infinity, so they apply to pure translations as well as rotations.

The Limit Surface

The form of equations 6.24 and 6.25 suggests a functional relationship between the slider's rotation center and the resulting frictional force. However, if we allow non-zero support force at a discrete point, then these equations are undefined for rotations about the support point. For this reason the relation between slider motion and frictional force cannot generally be described as a function. Fortunately there is an elegant description of the motion–force mapping: the *limit surface* introduced by Goyal, Ruina, and Papadopoulos (1991).

To develop the limit surface, we first consider sliding of a single particle. Let $\mathbf{v}$ be the velocity of the particle, and let $\mathbf{f}$ be the frictional force applied *by the particle to the support surface*. Note that this convention is the opposite of our usual convention, and corresponds to a sign change on the force $\mathbf{f}$. We will use the term *frictional load* when referring to the frictional force applied by the slider to the support surface.

Friction

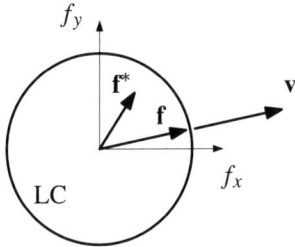

Figure 6.13
Limit curve for a point slider. Adapted from (Goyal et al., 1991).

For our point slider, we can express Coulomb's law as follows:

slip: $\mathbf{f} \parallel \mathbf{v}$, and $|\mathbf{f}| = \mu f_n$, where μ is the coefficient of friction, and f_n is the support force.

stick: $|\mathbf{f}| \leq \mu f_n$.

We can develop an equivalent graphical representation of Coulomb's law. Consider the set of all frictional loads that can be applied by the point slider. This set is a disc comprising all forces acting through the point with magnitude no greater than μf_n. This set is bounded by a circle of radius μf_n at the origin of force space, which we define to be the *limit curve* LC (see figure 6.13). Now we can say that the frictional load $\mathbf{f}$ must satisfy the *maximum power inequality*:

$$\forall_{\mathbf{f}^* \in \text{LC}} \ (\mathbf{f} - \mathbf{f}^*) \cdot \mathbf{v} \geq 0 \tag{6.26}$$

In other words, the motion $\mathbf{v}$ yields a load that is extremal in the $\mathbf{v}$ direction.

More significantly, we note that when slip occurs the load $\mathbf{f}$ is on the limit curve, and the motion $\mathbf{v}$ is *normal* to the limit curve at $\mathbf{f}$.

Now we consider sliders with extended support. Let $\mathbf{r}$ vary over the support region, and construct a limit curve $\text{LC}(\mathbf{r})$ at each point $\mathbf{r}$, so that at each point the maximum power inequality holds:

$$\forall_{\mathbf{f}^*(\mathbf{r}) \in \text{LC}(\mathbf{r})} \ (\mathbf{f}(\mathbf{r}) - \mathbf{f}^*(\mathbf{r})) \cdot \mathbf{v}(\mathbf{r}) \geq 0 \tag{6.27}$$

Now let $\mathbf{p}$ be the total frictional load wrench

$$\mathbf{p} = \begin{pmatrix} f_x \\ f_y \\ n_{0z} \end{pmatrix} = \sum_{\mathbf{r}} \begin{pmatrix} f_x(\mathbf{r}) \\ f_y(\mathbf{r}) \\ \mathbf{r} \times \mathbf{f}(\mathbf{r}) \end{pmatrix} \tag{6.28}$$

and let **q** be the velocity twist

$$\mathbf{q} = \begin{pmatrix} v_{0x} \\ v_{0y} \\ \omega_z \end{pmatrix} \tag{6.29}$$

Now, for some given motion **q**, let **f**(**r**) be a distribution of frictional loads satisfying Coulomb's law, and let **f***(**r**) be some arbitrary distribution, satisfying only the constraint that at each **r**, **f***(**r**) is in the corresponding limit curve:

$$\forall_r \ \mathbf{f}^*(\mathbf{r}) \in \mathrm{LC}(\mathbf{r}) \tag{6.30}$$

Let **p** and **p*** be the total frictional load wrench for **f**(**r**) and **f***(**r**) respectively. Now, we can describe the power dissipated by **f**(**r**) in either of two ways, yielding the equation

$$\mathbf{p} \cdot \mathbf{q} = \sum_r \mathbf{f}(\mathbf{r}) \cdot \mathbf{v}(\mathbf{r}) \tag{6.31}$$

Similarly we can write

$$\mathbf{p}^* \cdot \mathbf{q} = \sum_r \mathbf{f}^*(\mathbf{r}) \cdot \mathbf{v}(\mathbf{r}) \tag{6.32}$$

Taking the difference yields

$$(\mathbf{p} - \mathbf{p}^*) \cdot \mathbf{q} = \sum_r (\mathbf{f}(\mathbf{r}) - \mathbf{f}^*(\mathbf{r})) \cdot \mathbf{v}(\mathbf{r}) \tag{6.33}$$

Since the maximum power inequality must be satisfied at every point **r**, every term in the sum on the right hand side is non-negative. Thus we obtain a maximum power inequality for the total frictional load wrench:

$$(\mathbf{p} - \mathbf{p}^*) \cdot \mathbf{q} \geq 0 \tag{6.34}$$

To summarize, to find the true frictional load, we can start with the set of all load distributions satisfying the constraint that at each point **r** the magnitude of the load must be no greater than $\mu f_n(\mathbf{r})$, and then choose a distribution that yields maximum power.

When the slider is not moving, any load distribution **f***(**r**) is possible, subject only to the constraint that at each point the magnitude of the load must be no greater than $\mu f_n(\mathbf{r})$. Form the set of all possible total frictional load wrenches **p***, and define the *limit surface* to be the surface of this set. Then we can summarize the maximum power inequality by stating that the frictional load wrench during slip yields maximum power over all wrenches in the limit surface. It follows that during slip the total frictional load wrench **p** lies on the limit surface, and the velocity twist **q** is normal to the limit surface at **p**.

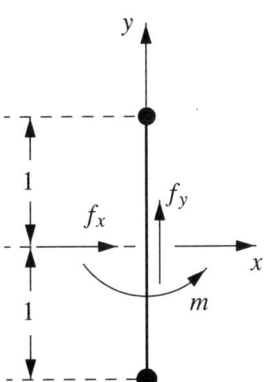

Figure 6.14
Sliding barbell. Adapted from (Goyal et al., 1991).

We state several properties of the limit surface without proof. The limit surface is closed, convex, and encloses the origin of wrench space. It is symmetric when reflected through the origin. Its orthogonal projection onto the f_x, f_y plane is a circle of radius $\sum \mu f_n$.

If the pressure distribution is everywhere finite, i.e. with no discrete points of support, then the limit surface is strictly convex, and the velocity twist to frictional load wrench mapping is one-to-one.

The more interesting cases involve discrete support points. If there are such points, then there are flat facets on the limit surface. At such a facet, several different loads give rise to the same motion—rotation about the discrete support point.

An even more interesting case arises when the support region R degenerates to a line or a subset of a line. In this case the limit surface is no longer smooth. At a vertex of the limit surface several different motions can produce the same frictional load. This corresponds to those motions with rotation centers collinear with all points of support.

The limit surface has uses that go well beyond what can be described here. It applies to some non-isotropic friction laws, such as ice skates or ratchet wheels. It yields insights into the dynamic motion of sliders, and, as we shall see it provides insights into the mechanics of quasistatic manipulation.

EXAMPLE

Figure 6.14 shows a planar slider with just two points of support, a *barbell*. We assume the barbell's weight is evenly divided between the two support points. Figure 6.15 shows the corresponding limit surface. It was constructed by the following steps:

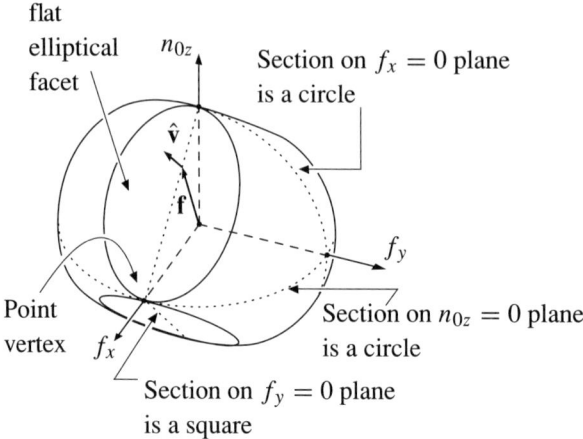

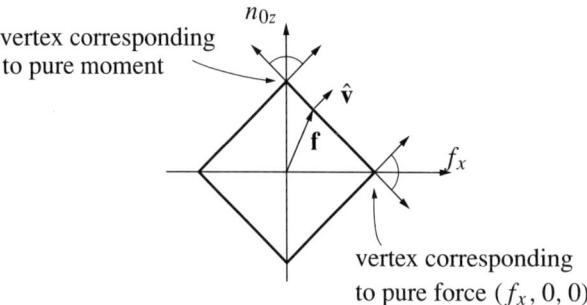

Figure 6.15
Barbell limit surface. Adapted from (Goyal et al., 1991).

1. Construct the limit surface LS_a comprising all force loads arising at support point a. If a were at the origin this would be a disc in the $n_{0z} = 0$ plane. But since a is not at the origin, LS_a is an elliptical disc in the $n_{0z} - f_x = 0$ plane.

2. Similarly, construct the limit surface LS_b. It also is an elliptical disc, this time in the $n_{0z} + f_x = 0$ plane.

3. The desired limit surface is the Minkowski sum of LS_a and LS_b. In other words it is the set $\{\mathbf{w}_a + \mathbf{w}_b \mid \mathbf{w}_a \in LS_a, \mathbf{w}_b \in LS_b\}$.

The barbell's limit surface illustrates many of the properties of limit surfaces. There are four flat facets, where the frictional load may vary while the normal remains stationary.

Friction

Figure 6.16
Two block in corner problems for exercise 6.2.

This implies many different loads mapping to a single motion, which occurs when the barbell rotates about one of the support points. There are four such facets, one for each of two possible rotation directions about each of two different support points.

There are also four vertices, where the frictional load is stationary as the normal may vary. This implies many different motions mapping to a single frictional load, which occurs for a rotation about a point collinear with the two support points.

Elsewhere the limit surface is smooth and strictly convex, so the load–motion mapping is one-to-one.

6.7 Bibliographic notes

Many engineering mechanics texts provide good introductions to Coulomb friction. Gillmor (1971) and Truesdell (1968) provide some interesting historical notes on Coulomb, Amontons, and da Vinci. Simunovic (1975) was the first to analyze peg insertion using friction cones. Erdmann (1984) was the first to construct composite friction cones in wrench space. Prescott (1923) and MacMillan (1936) developed expressions for force and moment of planar sliding, and introduced the center of friction. The particular treatment of planar sliding is taken from (Mason, 1986). The limit surface is taken from (Goyal, 1989; Goyal, Ruina, and Papadopoulos, 1991). See (Howe and Cutkosky, 1996) for experimental evaluation, application, and approximations related to the limit surface.

Exercises

Exercise 6.1: Analyze the pipe clamp using the moment labeling method, and the force dual method: find the possible resultants of the contact forces, and characterize the set of load forces that would be balanced.

Exercise 6.2: Use the moment labeling and the force dual methods to analyze each of the problems in figure 6.16. A block is in the corner of a fixed tray, and you are to identify the

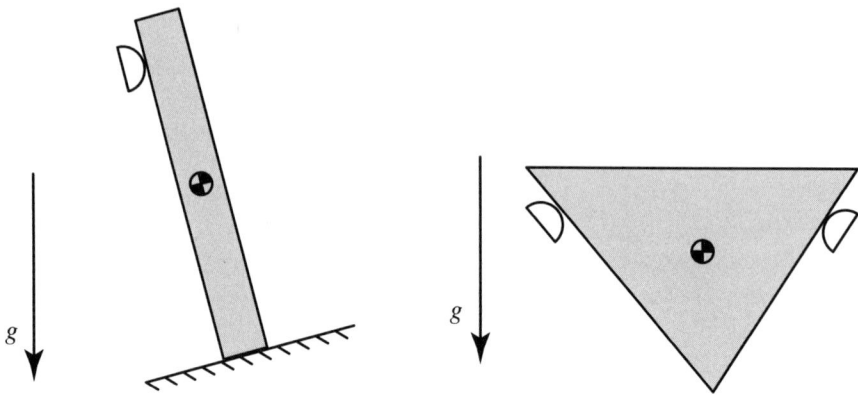

Figure 6.17
Two equilibrium problems. Exercises 6.3 and 6.4.

set of all wrenches that can be applied to the block by the tray edges. The coefficient of friction is one. If you get confused, make the problem simpler by taking a smaller coefficient of friction, and then think about the solution in the limit as μ approaches one. Do not forget to look for intersections at infinity.

Exercise 6.3: The planar rectangle in figure 6.17(a) is resting on a tipped table and one additional contact, in a gravitational field. Is there a static equilibrium? Use moment labeling to analyze it. If it is unstable, find a location for the finger that would stabilize it, or show that no such location exists.

Exercise 6.4: The triangle of figure 6.17(b) is resting on two fingers. Given that there is a static equilibrium, find a lower bound for the coefficient of friction. The only forces acting are the contact forces and gravity.

Exercise 6.5: In exercise 5.7 you constructed the force dual representation of all contact normals of a triangle (the zigzag locus). That corresponds to the set of all wrenches that can be applied by a single frictionless contact. For this exercise, construct the force dual representation of all wrenches that can be applied to the same triangle by a single contact, with coefficient of friction 0.25.

Exercise 6.6: Howe and Cutkosky (1996) observed that in many cases an ellipsoid is a good approximation to the limit surface for a planar slider. Suppose that an ellipsoidal

Friction

limit surface has the form

$$\frac{f_x^2}{a^2} + \frac{f_y^2}{b^2} + \frac{n_{0z}^2}{c^2} = 1$$

(a) Prove that $a = b$.
(b) Find a simple closed form expression giving any differential twist (v_x, v_y, ω_z) as a function of f_x, f_y, n_{0z}, a, and c.
(c) Find a simple closed form expression giving the frictional load (f_x, f_y, n_{0z}) as a function of v_x, v_y, ω_z, a, and c.

Exercise 6.7: Consider the barbell of figure 6.14, with weight 2 Newtons evenly distributed between the two contact points, and coefficient of friction $\mu = 1$. Redraw figure 6.15 with correct numerical labels along the axes, indicating the maximum values of f_x, f_y, and n_{0z}. Calculate the frictional load for each differential twist below. If several frictional loads map to the given twist, give a concise and precise description of the set.
(a) $(v_x, v_y, \omega_z) = (1, 0, 0)$
(b) $(v_x, v_y, \omega_z) = (0, 1, 0)$
(c) $(v_x, v_y, \omega_z) = (0, 0, 1)$
(d) $(v_x, v_y, \omega_z) = (1, 0, 1)$

Exercise 6.8: Consider a dinner plate whose entire weight w is evenly distributed on a ring at radius 1, centered on the origin. The coefficient of friction is 0.25. Because of the symmetry in the plate, the limit surface has circular symmetry and is completely characterized by its intersection with the f_x-n_{0z} plane.
(a) Rewrite equations 6.24 and 6.25 to give the frictional load (i.e. take care of the sign change) and use a differential element of length dl rather than area dA.
(b) Plot the limit surface's intersection with the f_x-n_{0z} plane. You can generate such a plot by numerically integrating the equations obtained above, with $v_y = 0$ and varying v_x/ω.

Exercise 6.9: We noted earlier the propensity for refrigerators to rotate about their feet. Use the nature of the limit surface to explain this phenomenon—i.e. explain why planar sliders like to rotate about points of infinite pressure.

7 Quasistatic Manipulation

This chapter explores several different manipulation tasks: grasping and fixturing, pushing, parts orienting, and mechanical assembly. We analyze each task using an approach known as *quasistatic analysis*, meaning that we look for a balance among contact forces, gravitation, and other applied forces, while neglecting inertial forces. This approach can be very accurate with the speeds and scale of objects often encountered in manipulation tasks, but will fail when dynamic forces come into play.

7.1 Grasping and fixturing

Grasping and fixturing are two variations on the most fundamental problem in manipulation: how to immobilize an object. To grasp is to immobilize relative to the hand; to fixture is to immobilize absolutely. There are other important aspects to each problem, but we will focus on how to immobilize objects.

Some of our most common tools are general-purpose fixtures. For example a table has a flat horizontal surface, which with the help of gravitation and friction serves to immobilize a wide variety of objects. Also, our most common objects are designed for fixturing. For example, the most common pencils are hexagonal in cross section, presumably to keep them from rolling off of tables.

Immobilizing objects is a topic that is revisited throughout the text. In chapter 2 we introduced Reuleaux's method to analyze kinematic constraint. In chapter 3 we developed inequalities in the object twist screw coordinates using contact screws. In chapter 5 we introduced moment labeling and force dual methods. In this section we bring all these methods together and address the issue in depth.

Let's begin with some definitions:

DEFINITION 7.1: **Force closure**: the contacts can apply an arbitrary wrench to the object.

This is consistent with our earlier definition, which defined force closure as being when the possible wrenches are all of wrench space.

DEFINITION 7.2: **Form closure**: the object is at an isolated point in configuration space.

In other words, every nearby configuration results in a collision. We include the definition of form closure for completeness. A fuller treatment of form closure is outside the scope of the text. Rather, we will only deal with first order form closure:

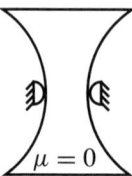
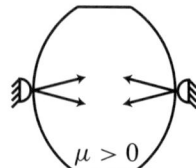

Form closure
does not imply
force closure

Force closure
does not imply
form closure

Figure 7.1

DEFINITION 7.3: **First order form closure**: Every nonzero velocity twist is contrary to some contact screw.

In other words, no motions are possible even if we approximate each contact constraint by a velocity constraint.

DEFINITION 7.4: **Equilibrium**: the contacts can balance the object's weight and other external forces.

We will not provide a formal definition of *stability*. It is a term that is applied to many different approaches. Consider, for example, the dynamic system comprising the object, a dexterous hand, and the control system that determines the finger joint torques. *Stability* would address the asymptotic properties of this dynamic system, which is clearly beyond the scope of this chapter, and in fact this entire book. However, in section 7.3 we apply quasistatic methods to address stability in the context of pushing.

What is the relation of force closure, form closure, and first order form closure? The first thing to notice is that first order form closure is equivalent to frictionless force closure. In both cases we consider whether the contact normals positively span wrench space. It follows easily that form closure is stricter than first order form closure and frictionless force closure.

The relationship between form closure and force closure is more interesting. Figure 7.1 shows by example that form closure does not imply force closure, and that force closure does not imply form closure.

For any standard we choose, there are three issues to consider:

1. *Analysis*. Given an object, a set of contacts, and possibly other information, determine whether closure applies.

2. *Existence*. Given an object, and possibly some constraints on the allowable contacts, does a set of contacts exist to produce closure?

3. *Synthesis*. Given an object, and possibly some constraints on the allowable contacts, find a suitable set of contacts.

We have explored the analysis question in earlier chapters. For force closure we look at the positive linear span of the friction cones (section 6.4) and check whether it is all of wrench space. For first order form closure we examine the positive linear span of the contact normals (section 5.3).

How do we address the existence issue? Recall the "zigzag locus", which gives the set of *all* contact normals for a given object. We can form the positive linear span of all these contact normals to answer the existence question for first order form closure or frictionless force closure. For frictional force closure we would take the positive linear span of all possible friction cones (exercise 6.5). Existence is just the question of whether the resulting convex cone exhausts all of wrench space.

Are there any shapes that do *not* have force closure grasps? There are a few, which can be described by looking at the lower pairs of figure 2.21. Since we are only interested in bounded shapes, the only pairs that qualify are revolute and spherical.

THEOREM 7.1: For any bounded shape that is not a surface of revolution, a force closure (or first order form closure) grasp exists.

For a proof, see (Mishra et al., 1987).

Synthesis

The most challenging issue is synthesis: given an object, how do we construct a grasp? We begin with a very simple algorithm. We will consider a finger to be redundant if it can be deleted without reducing the positive linear span of all the fingers.

```
procedure GRASP
    put fingers "everywhere"
    while redundant finger exists
        delete any redundant finger
```

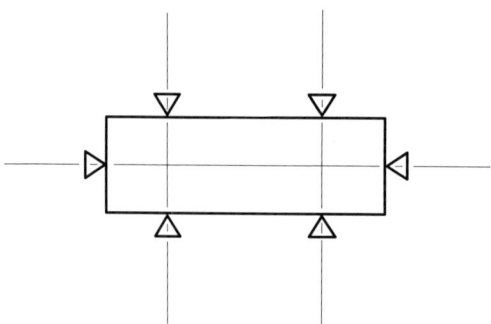

Figure 7.2
An example of six contacts, none of which can be eliminated with frictionless force closure.

The idea is to sample the boundary of the object very densely, so that we begin with a large but finite set of contacts. Unless the object is a frictionless surface of revolution, the object will be in force closure and most of the contacts will be redundant. Now we discard redundant contacts one at a time until no redundant contacts are left.

It is clear that, unless we start with a surface of revolution, this algorithm yields a force closure grasp. The question is: how good is the resulting grasp? And that leads us immediately to the more basic question: how do we measure the quality of a grasp? Here we will consider only how many contacts are produced. We already know (theorem 5.6) that a frictionless force closure grasp requires at least 4 contacts in the plane, 7 contacts in three space. Generally we might expect GRASP to terminate with just 4 or 7 contacts, but in some unfortunate cases the algorithm can terminate with more.

THEOREM 7.2 STEINITZ'S THEOREM: Let X be a set of points in $\mathbf{R}^d$, with some point p in the interior of the convex hull of X. Then there is some subset Y of X, with $2d$ points or less, such that p is in the interior of the convex hull of Y.

We can apply Steinitz's theorem in wrench space. If we start with a set of wrenches that positively span wrench space, then the origin is in the convex hull of the wrenches (theorem 5.5). By Steinitz's theorem, there is a subset of those wrenches, numbering $2d$ or less, which still positively span wrench space.

THEOREM 7.3: For any surface not a surface of revolution, GRASP yields a grasp with at most 6 fingers in the plane, at most 12 fingers in three space.

Six fingers seems like a lot for a planar grasp. Figure 7.2 shows such an unfortunate grasp. (Figure 5.19 enumerates all similar examples.) Even though we know a four contact grasp exists for a rectangle, GRASP has terminated with a six contact grasp. None of the six contacts can be deleted without losing closure. However, if we perturb the six contacts to eliminate geometrical coincidences, the grasp could be reduced to four contacts.

We have seen that idealized grasps and fixtures, employing fixed contacts on a rigid object, have a tidy theoretical foundation based on wrench space. But we have barely scratched the surface. Grasp and fixture planning is an active research area, with lots of interesting problems. The bibliographic notes at the end of the chapter describe a few of these.

7.2 Pushing

Pushing involves a planar slider being moved by contact with a pusher whose motion is given. There are two frictional contacts to address. The first is the frictional planar contact between the slider and the horizontal surface surface, so that the analysis of section 6.6 can be applied. The second frictional contact is between the slider and the pusher, which we will assume also follows Coulomb's law. There may also be frictional contacts with other objects in the scene but we will focus on simpler problems. In this section we assume a single point contact with the slider. In the next section we explore the problem of pushing a slider with edge-to-edge contact.

Pushing is more common than you might realize. Here are some examples:

- To pick up objects that are too small or too numerous to grasp easily, you can scoop them off the edge of a table into your other hand;
- To move objects that are too bulky to grasp, as when rearranging the furniture, you can push them;
- Manufacturing automation systems make frequent use of pushing. Often a conveyor belt, in conjunction with guides, is used to move objects through such a system.

Pushing can be a good way to reduce or eliminate uncertainty in the state of the task. Figure 7.3 shows some examples: a fence across a conveyor belt to orient boxes; and a grasping operation that orients and centers the grasped object. Both of these examples depend largely on the mechanics of pushing.

Which way does it turn?

As observed in section 6.6, the motion of an object being pushed through a single point of contact is often indeterminate. Despite this indeterminacy, it is sometimes possible to

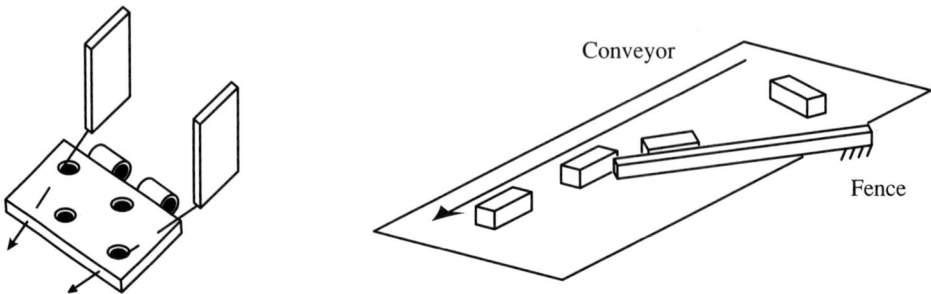

Figure 7.3
Examples of pushing: orienting and locating an object during a grasp; orienting boxes by a fence suspended just above a conveyor.

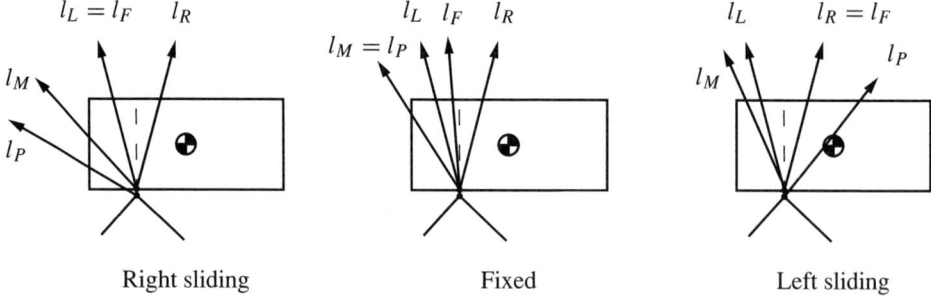

Figure 7.4
Possible relations of rays: the line of pushing l_P, the line of motion l_M, the left and right edges of the friction cone l_L and l_R, and the line of force l_F.

predict qualitative features of the motion, including whether the object will rotate and in which direction.

DEFINITION 7.5: We define the following directed lines:

- The **line of pushing** passes through the contact point, in the direction of the pusher velocity.
- The **line of motion** passes through the contact point, in the direction of the object velocity.

Figure 7.4 shows the line of pushing l_P, and the line of motion l_M, for a few different examples. Also plotted are the left and right edges of the friction cone, l_L and l_R

Quasistatic Manipulation

respectively, and the line of force l_F. Note that the friction cone and the line of pushing are determined by the contact geometry, the pusher motion, and the coefficient of friction. However, the line of motion and line of force are not so easily predicted. Constraints on the line of motion and line of force can be stated, for each contact mode.

- *Separation.* The object just sits there. There is no line of force and no line of motion.
- *Fixed.* The line of motion coincides with the line of pushing: $l_P = l_M$, and the line of force falls between the left and right edges of the friction cone.
- *Left sliding.* The line of motion falls to the left of the line of pushing, and the line of force coincides with the right edge of the friction cone: $l_F = l_R$.
- *Right sliding.* The line of motion falls to the right of the line of pushing, and the line of force coincides with the left edge of the friction cone: $l_F = l_L$.

(One of the tricky bits of these problems is to remember which is left and which is right. We adopt the *pusher* as the reference since its motion is given, so "left sliding" means the slider is sliding to the left relative to the pusher. The easiest way to remember is to recall that Coulomb friction opposes the motion—left sliding means right edge of friction cone, and vice versa.)

The main result of this section is to show that the direction of rotation can be determined by the relation of the center of friction to the three lines: l_P, l_L, and l_R. But we have to get there in a roundabout way, by exploring the relation to the other two lines: l_M and l_F. First it will help to introduce some terminology. We will say a directed line l *dictates* the rotation direction, if the sign of the rotation must agree with the sign of the moment of l, with respect to the center of friction. Similarly, we will say that three lines *vote* on the rotation direction, if the sign of the rotation must agree with a majority of the relevant moments.

THEOREM 7.4: For quasistatic pushing of a rigid body in the plane, with uniform coefficient of friction, the line of motion dictates the rotation direction.

Proof Choose the origin at the contact point, and choose the coordinate frame with the y-axis coincident with the line of motion l_M. Then the rotation center must lie on the x-axis, and we can write $\mathbf{r}_{IC} = (x_{IC}, 0)^T$. Note that x_{IC} can take on any value on the plus or minus x-axis, and can be on the line at infinity, but cannot be zero. Now, consider the total moment of frictional forces, with respect to the origin, as a function of x_{IC}.

$$m_f(x_{IC}) = -\mu \operatorname{sgn}(\dot\theta) \int_R \mathbf{r} \cdot \frac{\mathbf{r} - \mathbf{r}_{IC}}{|\mathbf{r} - \mathbf{r}_{IC}|} p(\mathbf{r})\, dA \qquad (7.1)$$

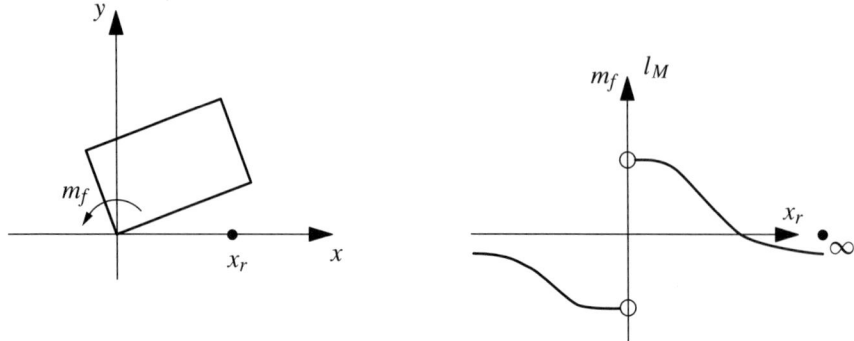

Figure 7.5
Coordinate frame for proof of theorem 7.4 has the y-axis along the line of motion l_M. The proof is obtained by analyzing the frictional moment as a function of the IC location.

The solution of our problem is given by finding a value of x_{IC} that yields a balance between the pushing force and the frictional forces. Since the pushing force acts through the origin, the moment at the origin is zero. So the solution x_{IC} is just the root of the equation

$$m_f(x_{IC}) = 0 \tag{7.2}$$

This root depends on the pressure distribution $p(\mathbf{r})$, which is indeterminate. However, the function $m_f(x_{IC})$ has a structure that permits us to prove the theorem, without requiring a specific solution for x_{IC}. The function $m_f(x_{IC})$ varies continuously as x_{IC} varies from small positive values, up through infinity, and approaching small negative values. We can take the derivative

$$\frac{d}{dx_{IC}} m_f(x_{IC}) = -\mu \, \text{sgn}(\dot{\theta}) \int_R \mathbf{r} \cdot \frac{d}{dx_{IC}} \frac{\mathbf{r} - \mathbf{r}_{IC}}{|\mathbf{r} - \mathbf{r}_{IC}|} p(\mathbf{r}) \, dA \tag{7.3}$$

and simplify to obtain

$$\frac{d}{dx_{IC}} m_f(x_{IC}) = -\mu \, |x_{IC}| \int_R \frac{y^2}{|\mathbf{r} - \mathbf{r}_{IC}|^3} p(\mathbf{r}) \, dA \tag{7.4}$$

where $\mathbf{r} = (x, y)^T$. Notice that the integrand is positive, so that the derivative of $m_f(x_{IC})$ is always negative—the function is monotonic decreasing.

We can also take limits of $m_f(x_{IC})$ as x_{IC} approaches zero from above and from below:

$$m_f(0+) = \mu \int_R |\mathbf{r}| \, p(\mathbf{r}) \, dA \tag{7.5}$$

$$m_f(0-) = -\mu \int_R |\mathbf{r}| \, p(\mathbf{r}) \, dA \tag{7.6}$$

Finally, we can determine the value of m_f where $x_{IC} = \infty$. This is the case of pure translation. Applying theorem 6.1 we have

$$m_f(\infty) = -\mu \int_R x \, p(\mathbf{r}) \, dA \tag{7.7}$$

$$= -\mu f_0 x_0 \tag{7.8}$$

This is enough information to determine whether the solution x_{IC} is on the positive x-axis, the negative x-axis, or at infinity. Suppose the line of motion makes a positive moment with respect to the center of friction. Then $x_0 < 0$ and $m_f(\infty) > 0$. By the intermediate value theorem, the root of $m_f(x_{IC})$ has to fall in the negative x-axis, so the rotation must be positive. Similarly, if the line of motion makes a negative moment with respect to the center of friction, then the rotation must be negative. And if the line of motion passes through the center of friction, a pure translation takes place. ∎

Presumably the reader will at this point be wondering why we cannot solve the problem with a simple force balance. In fact the method is, at its foundation, a force balance, but it is not straightforward because the pressure distribution is not known. The reader might also be wondering whether some variational method might be employed—whether the solution might minimize work. In fact, it is possible to show that the solution minimizes the energy dissipated in frictional sliding, as discussed in the bibliographic notes at the end of the chapter. For our present purposes we will apply the more straightforward approach.

THEOREM 7.5: For quasistatic pushing of a rigid body in the plane, with uniform coefficient of friction, the line of force dictates the rotation direction.

For a formal proof see (Mason, 1986). Alternatively, consider the Limit Surface introduced in section 6.6. If we choose the center of friction as the origin, then the limit surface intersects the horizontal (f_x-f_y) plane in a circle, where the limit surface normal is also in the horizontal (v_x-v_y) plane. By the convexity of the limit surface, the normals in the upper half point upward and in the lower half point downward. Positive rotations correspond to frictional loads with positive moments, and negative rotations correspond to frictional loads with negative moments. This is equivalent to the statement of the theorem.

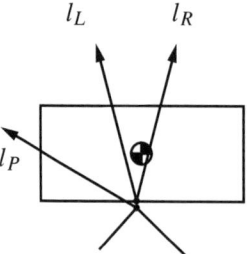

Figure 7.6
The case analyzed in the proof of theorem 7.6.

Now we can prove the main result:

THEOREM 7.6: For quasistatic pushing of a rigid body in the plane with uniform coefficient of friction, the rotation direction is determined by a vote of three directed lines: the line of pushing and the left and right edges of the friction cone.

Proof The simplest case is when the two edges of the friction cone agree, which occurs when the center of friction is outside the friction cone. Since the line of force falls inside the friction cone, and dictates the rotation direction (theorem 7.5), the theorem follows easily.

The more interesting case is when the edges of the friction cone disagree. We will consider one such case shown in figure 7.6; the other cases are similar. In the figure, l_L votes $-$, l_R votes $+$, and l_P votes $-$. The majority is $-$. We will assume a positive rotation, and derive a contradiction. Since l_P votes $-$, it is to the left of the center of friction $\mathbf{r}_0$. Since a positive rotation occurs, theorem 7.4 says that l_M is to the right of $\mathbf{r}_0$. That means l_M is to the right of l_P, giving the right-sliding contact mode. Right sliding requires a line of force l_F on the left edge l_L of the friction cone. Theorem 7.5 says the line of force dictates the rotation direction—it must be negative, contradicting our assumption of positive rotation. We conclude the rotation has to be negative or zero, but translation is easily excluded. ∎

Theorem 7.6 is particularly useful in analyzing tasks such as the grasping and orienting tasks of figure 7.3. We will return to parts orienting tasks in section 7.4. First, we turn to the question of stably pushing an object with edge contact.

Figure 7.7
Stable pushing of a box by a mobile robot.

7.3 Stable pushing

The previous section developed some of the basic mechanics of pushing. In this section we develop some additional theory and planning techniques for pushing an object to a desired location.

Consider a simple pushing operation, such as the one illustrated in figure 7.7. To maintain control of the box, the robot uses an edge-to-edge contact, and it chooses motions that will result in *stable pushing*, meaning that the edge-to-edge contact is maintained. Planning such an operation involves two problems: first we must identify the stable pushing motions, and second we must find a collision-free path using only those motions. We assume that the start and goal configurations are given, that the slider shape and center of friction is known, and that a lower bound on the coefficient of friction at the pushing contact is given. The problem is to find a path for the pusher and slider such that

- the slider remains fixed with respect to the pusher during the motion (stability),
- the path is collision free from the start to the goal,
- any additional pusher motion constraints are observed. For example, if the pusher is a wheeled mobile robot we have to use motions that do not require sideways motions of the wheels.

We will employ the nonholonomic planner NHP described in chapter 4, so as input we also need to specify a cost function.

The main challenge is stability. The mobile robot must employ motions that avoid the two problems shown in figure 7.8. If the robot moves the wrong way, the slider can slip off the pusher, or it can roll on the pusher. The question is, what robot motions will produce a stable push? To address that question we need to introduce a few results that bound the possible instantaneous rotation centers of an object being pushed. We will develop three bounds, which when combined suffice to produce stable pushing motions.

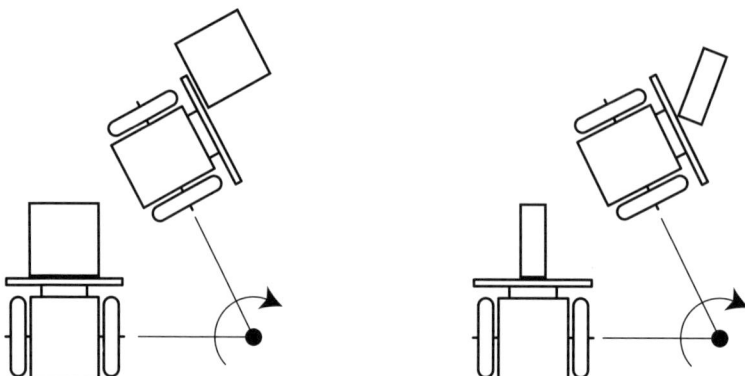

Figure 7.8
Two failure modes: the box can slide or roll off the pusher.

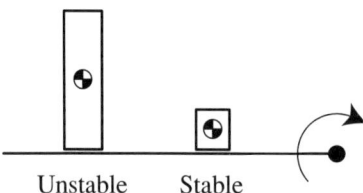

Figure 7.9
Different bodies can be stably turned at different rates.

Peshkin's bound

Some objects tend to turn quickly, and some do not (figure 7.9). Roughly speaking, an object with a broad pressure distribution turns slowly. Of course, the greater the moment arm of an applied force, the more an object tends to turn. Peshkin's bound addresses these two effects, relating an object's tendency to turn to the difference between the applied moment arm and the pressure distribution's radius.

To develop this bound, we proceed as follows:

1. Circumscribe a disk around the slider, centered at the center of friction. Figure 7.10 shows an example disk, and the applied wrench. Imagine that you generated every possible pressure distribution for the disk, and plotted the resulting instantaneous rotation center. Then the result is the locus of possible instantaneous centers consistent with the given applied wrench.

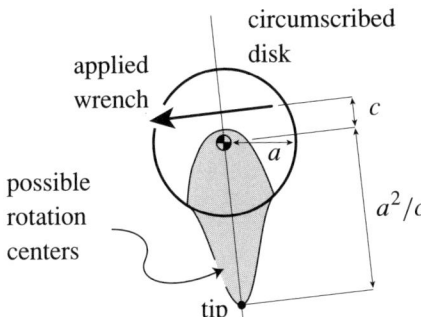

Figure 7.10
For the given applied wrench and a pressure distribution confined to the disk, the IC will be somewhere in the shaded region.

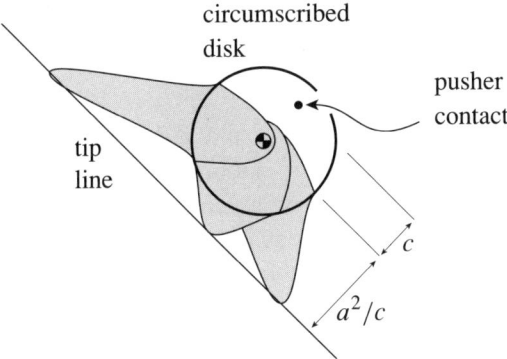

Figure 7.11
As we vary the angle of the line of force, the locus of possible ICs varies so that the tip sweeps out a tip line.

2. For a push with the applied wrench shown, the tip of the locus would yield the slowest rotation of the object. Let a be the radius of the disk, and let c be the moment arm of the applied wrench. Then the tip lies at a distance of a^2/c from the center of friction. An interesting and useful point: the mapping from wrench to tip is a variation on the familiar force-dual map. The force dual construction of figure 5.15 describes the behavior of the disk, if we select the unit of length to be the disk radius a.

3. Now suppose we are pushing the slider with a single point contact. We know that the applied wrench passes through the contact point, but we do not know in which direction. Applying the dual map to the contact point yields the tip line (figure 7.11).

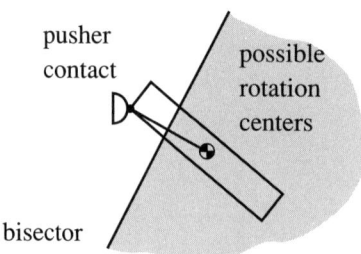

Figure 7.12
The bisector bound: no matter how the support pressure is distributed, the IC must fall in the shaded region.

This gives us a useful bound on the possible slider motions which we will call the *Peshkin bound*:

Given a slider circumscribed by a disk centered on the center of friction, the instantaneous center must fall inside the tip line, which is dual to the contact point.

(Actually the locus of instantaneous centers does bulge slightly outside the tip line, so the tip line should be viewed as a slightly fuzzy line.)

How do we know the Peshkin bound is a true bound? Peshkin studied *dipods*: pressure distributions involving just two points of support. The boundary of the instantaneous center loci in figures 7.10 and 7.11 are generated by dipods. Peshkin conjectured that *every* pressure distribution in the disk would yield an instantaneous center within the limits defined by the dipods, and he supported this conjecture by randomly generating lots of pressure distributions, none of which exceeded the limits defined by the dipods. Peshkin's conjecture remains unproven, but nobody has ever produced a counterexample. Thus the Peshkin bound comes with an interesting guarantee: if it fails, you can claim the honor of discovering the counterexample to Peshkin's conjecture.

The "bisector" bound

Figure 7.12 illustrates the *bisector bound*:

Given a contact point and a center of friction, construct the perpendicular bisector. The instantaneous center must lie in the delimited half-plane.

The bisector bound is parallel to Peshkin's tip line, so that together they define a strip which bounds the possible instantaneous centers.

How do we know the bisector bound is a true bound? In fact no proof has ever been published. Randy Brost and I used the bound in a paper of 1986, but did not include the proof. For some suggestions on the proof see exercise 7.6.

Quasistatic Manipulation

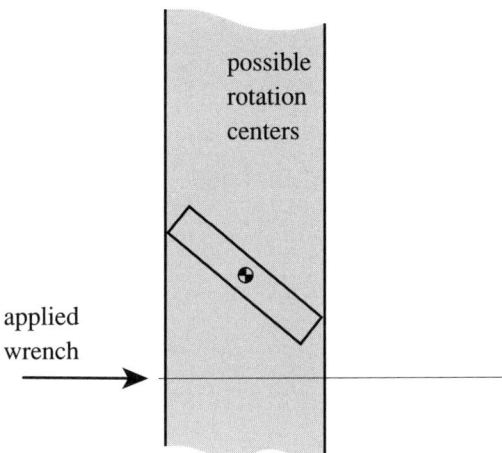

Figure 7.13
The vertical strip bound. No matter how the support pressure is distributed, the IC must fall in the shaded strip.

The "vertical strip" bound

Figure 7.13 illustrates another useful bound on instantaneous centers called the *vertical strip bound*:

> *The perpendicular projection of the IC onto the line of action must fall within the projection of the slider support region.*

How do we know the vertical strip bound is true? Suppose a wrench is applied to a planar slider. We choose a coordinate system with the x-axis along the line of action. Suppose the support region is bounded by a vertical strip $x_1 \leq x \leq x_2$. Then the vertical strip bound claims that the instantaneous center of rotation must be in the same strip. Consider the case of a negative rotation about a rotation center to the left of the vertical strip, as in figure 7.14. Then every possible support point will have a negative velocity in the y-direction and thus, by Coulomb's law, a positive component of force in the y-direction. Integrating over the entire support region has to give a positive component of force in the y-direction, which cannot balance the applied wrench. The other cases are similar.

Computing stable push motions

Now we consider how to use the three bounds (the Peshkin bound, the bisector bound, and the vertical strip bound) to find a set of pusher motions that will give us stable pushing.

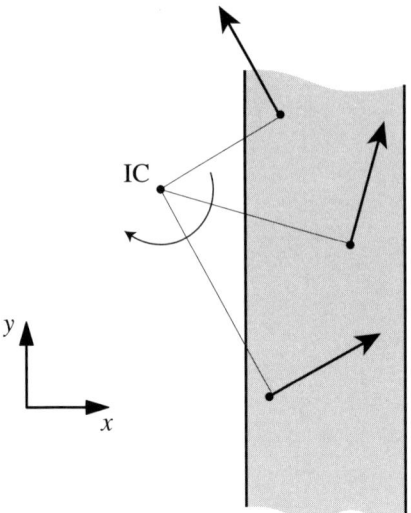

Figure 7.14
Proof of the vertical strip bound. An IC outside the shaded region yields a vertical component of force that cannot be balanced by the given wrench.

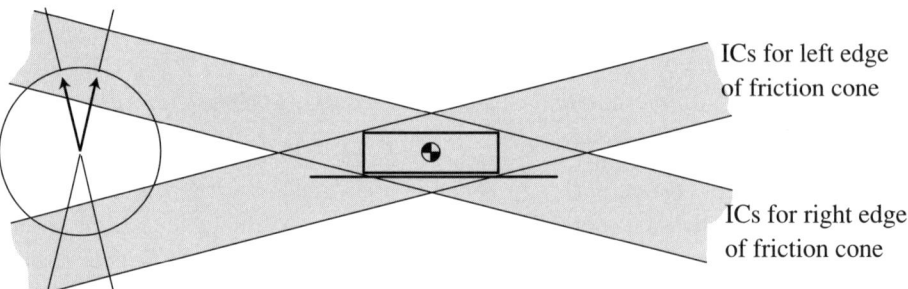

Figure 7.15

Consider the possible failure modes of figure 7.8. Note that if the slider slips on the pusher then the applied wrench must be on the left or right edge of the friction cone. Also note that if the slider rolls on the pusher, the applied wrench must be through the left or right vertex. We proceed by eliminating these possible failure modes.

First we can use the vertical strip bound to eliminate the slipping failure mode (figures 7.15 and 7.16). We consider the set of all wrenches with direction satisfying Coulomb's law. Slip corresponds to the subset of wrenches at either the left or right edge

Quasistatic Manipulation 159

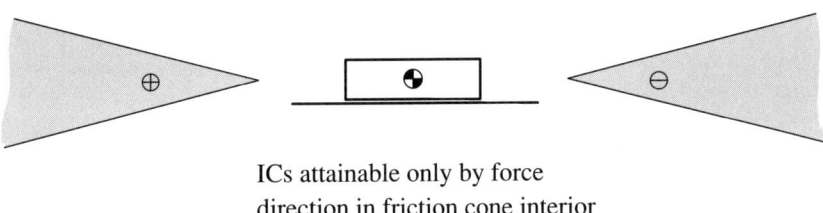

ICs attainable only by force
direction in friction cone interior

Figure 7.16

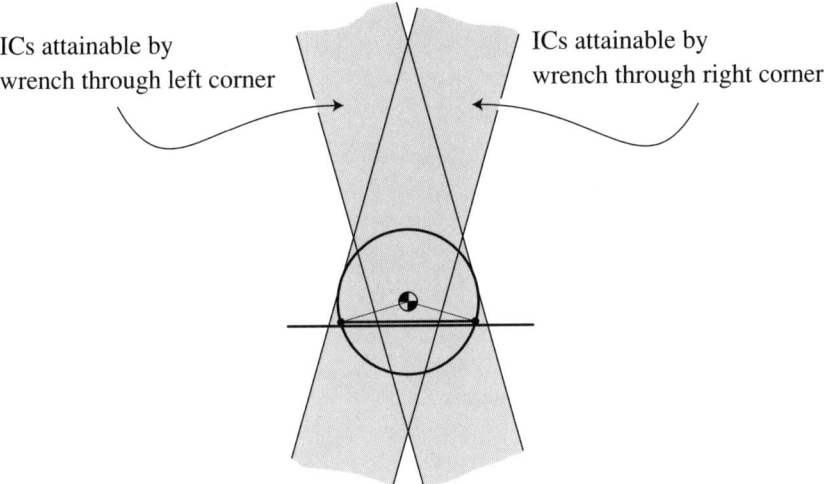

ICs attainable by
wrench through left corner

ICs attainable by
wrench through right corner

Figure 7.17

of the friction cone. Figure 7.15 shows the ICs that might possibly map to the left or right edge of the friction cone, and figure 7.16 shows the ICs that might possibly map to the interior of the friction cone, but definitely do not map to either edge of the friction cone.

Now we can use the Peshkin and bisector bounds to eliminate the rolling failure mode (figures 7.17 and 7.18). We consider the set of all wrenches with line of action through the contact line segment, regardless of the direction. Rolling corresponds to the subset of wrenches at either the left or right vertex of the contact line segment. Figure 7.17 shows the ICs that might possibly map to a line of action through the left or right vertex of the contact line segment. Figure 7.18 shows the ICs that might possibly map to a line of action through the interior of the contact line segment, but definitely do not map to either vertex.

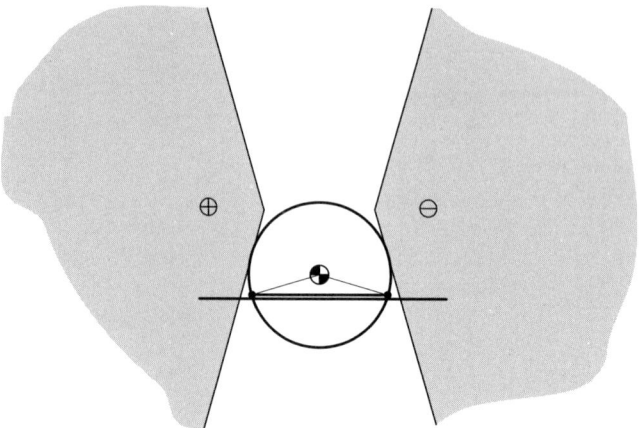

ICs attainable only by wrench
between the two corners

Figure 7.18

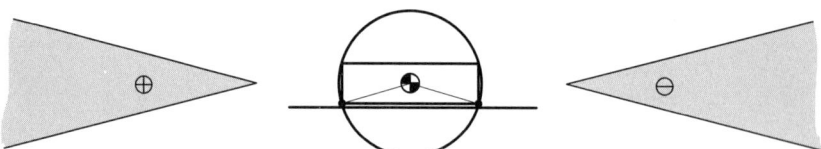

Figure 7.19
Combining the constraints of figures 7.16 and 7.18 gives a set of ICs yielding wrenches that act between the two corners and in the interior of the friction cone, so that neither the slipping nor rolling failure modes are possible.

If we intersect the ICs of figures 7.16 and 7.18 we obtain figure 7.19. These are the ICs that might map to a wrench that the pusher can apply to the slider during a stable push, but definitely cannot map to a failure mode. If the pusher uses one of these ICs, the slider has to move without slipping or rolling, i.e. it has to move with the same IC, giving a stable push.

The construction above can be stated more succinctly, albeit more abstractly. The pushing geometry and Coulomb's law define a polyhedral convex cone of possible pushing wrenches. The failure modes correspond to the boundary of that polyhedral convex cone, and the stable pushes correspond to the interior. We use the bounds to construct all ICs possibly mapping to the polyhedral convex cone, and then eliminate the ICs consistent

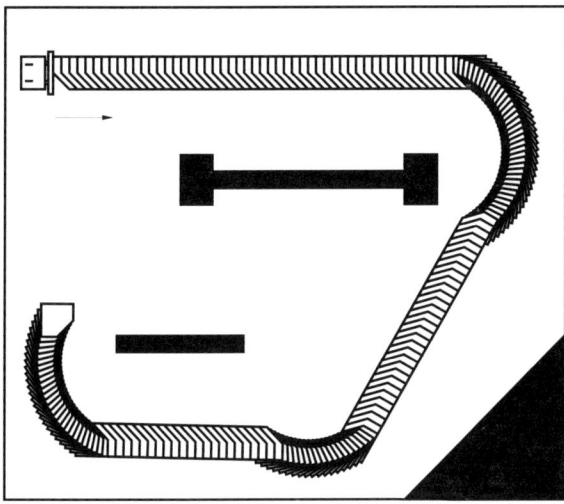

Figure 7.20
An automatically planned path for a mobile robot pushing a polygonal object. From (Lynch and Mason, 1996).

with the *boundary* of the polyhedral convex cone. Any remaining IC must map to the interior of the polyhedral convex cone, which is inconsistent with slipping or rolling.

We have to consider the possibility that we have been too conservative. It seems possible that we will eliminate *all* ICs. Perhaps for every IC we might consider, there is some pressure distribution mapping that IC to a failure mode. Fortunately that is not the case for suitable choices of pushing geometry. Suppose we choose a pushing edge so that there is a line of action through the interior of the pushing edge, through the center of friction, with direction that is interior to the friction cone. Then it is easy to show that a purely translational push in that direction is stable. Further, the construction above is guaranteed to return a set of ICs including that stable translational push as well as some ICs turning both left and right (Lynch and Mason, 1996).

Planning a stable pushing path

We now have a method for finding instantaneous centers that will produce stable pushes, at least in some simple cases. We can combine these constraints with any other applicable constraints, and then plan paths using the methods of chapter 4. For example, for the mobile robot pictured in figure 7.20 the rotation center must lie along the rear axle. Intersecting with the stable pushing constraints, we obtain a line segment in the oriented plane, corresponding to the feasible stable pushing rotation centers.

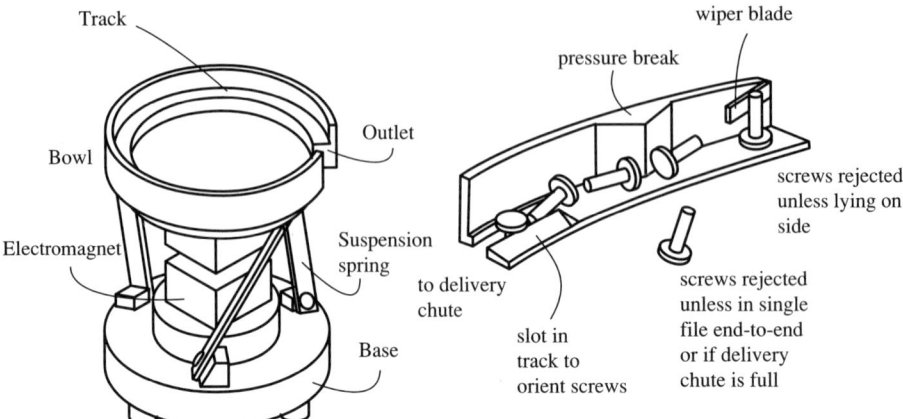

Figure 7.21
A vibratory bowl feeder uses a vibratory motion to move parts up a track past a sequence of obstacles. Adapted from (Boothroyd, 1992).

To apply NHP, we need to identify a small number of actions. In this case the choice is easy: we take the sharpest turning radius to the left, the sharpest turning radius to the right, and straight. With a suitable objective function, NHP constructs paths such as the one in figure 7.20.

7.4 Parts orienting

This section introduces the problem of parts orienting and feeding, and applies the theory of pushing to the problem.

Parts orienting is an important part of automation systems. Chapter 1 described one technique implemented by the Sony APOS system. Figure 7.21 shows a more common method: a *vibratory bowl feeder*. The parts are placed in the bottom of the bowl. A specially shaped vibration of the bowl causes the parts to climb a ramp that spirals up the inside wall. As the part climbs the ramp, it has to pass various stages which are designed to pass only single parts in the desired orientation.

The vibratory bowl feeder and the APOS machine provide an interesting contrast in *flexibility*: the ease with which a machine may be reconfigured for manufacture of a new product. For a bowl feeder, it typically takes a period of weeks or months to go from a part shape to a working feeder. For the APOS machine, it typically takes from one day to one week. In principle, robots can exhibit even greater flexibility. The ideal would be a system that requires no redesign or reconfiguration to orient a new part, but only a change in robot motion.

Quasistatic Manipulation

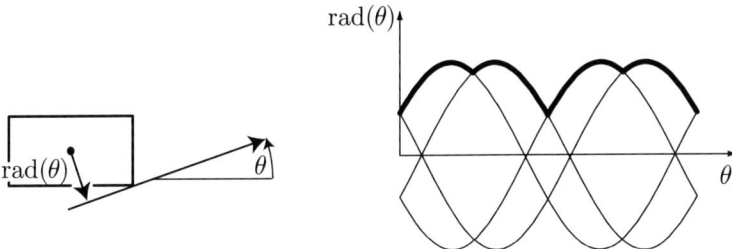

Figure 7.22
Notation and example of radius function.

Pushing is a good way to orient a part. With a flat support surface and a flat pusher, the same hardware can be used for a very broad variety of parts. This raises an important problem: to find a sequence of motions that will orient a given part. We will follow Goldberg's (1993) work to address a simplified version of the problem, subject to the following assumptions:

1. Shapes: The part is an isolated rigid planar polygon, on a planar support surface. The pusher is a flat rigid plate.
2. Forces: Contact forces are subject to Coulomb's law. The coefficient of friction with the support surface is uniform over the surface.
3. Motions: The pusher always moves along its own surface normal (*square* pushing). The part makes contact only with the face of the pusher. Each push proceeds until the part reaches a stable orientation.
4. Quasistatic: a balance of contact forces and gravity determines the object motion with sufficient accuracy.

The most awkward of these assumptions is that the push proceeds until the part reaches a stable orientation. In theory, that could require an arbitrarily long push. In practical situations this may or may not be a problem.

Radius function and push function

Square pushing (assumption 3 above) simplifies analysis of pushing. Recall the object rotation direction is determined by a vote of the line of pushing and the two friction cone edges (theorem 7.6). For square pushing the line of pushing is the contact normal, splitting the two friction cone edges, so the contact normal dictates the rotation direction.

The easiest way to see the result of square pushing is by the *radius function* of the pushed object. We place the origin of our coordinate system at the center of friction. Now consider a support line (contacting the boundary but not the interior) below the object, parallel to the x-axis. The distance from the origin to that support line is defined to be the *radius* at 0°. Now imagine that the support line rolls around the object in a counterclockwise direction. Let the angle of the support line be θ, and define the radius function rad(θ) to be the distance from the origin to the support line with direction θ.

Note the radius function is easily constructed as the maximum of a family of sinusoids. Each sinusoid corresponds to the radius function of a single vertex.

Suppose the support line in the above construction is the pusher. We fix the pusher orientation, and let the object rotate. Note that θ does *not* follow the usual convention for measuring object orientation. It gives the pusher orientation relative to the object, even though it is the object that moves. Equivalently, it is the angle from the pusher to the object measured *clockwise*, i.e. the negative of the usual convention.

Now we invoke theorem 7.6, and we ask the question: at what values of θ does the vote change? In the case of square pushing, the vote is dictated by the contact normal. So when does the center of friction switch from one side of the contact normal to the other? In either of two cases: when the center of friction coincides with the contact normal, which occurs at a maximum of the radius function; or when the object rolls from one vertex to another, which occurs at a minimum of the radius function. A maximum occurs only at the maximum of an individual sinusoid. Ordinarily a minimum occurs only at the intersection of two sinusoids, but see exercise 7.10 for an exception. For simplicity we will assume that the individual sinusoids do *not* intersect at any individual sinusoid's maximum. See exercise 7.17 for an example of what happens when we relax that assumption.

So, qualitatively, the radius function is like a potential function. Peaks are unstable equilibria, and valleys are stable equilibria. We can define a *push function* push(θ) (figure 7.23) to describe the effect of a square push: push(θ) returns the object orientation θ' resulting from a square push with initial object orientation θ. The push function is piecewise constant: push(θ) maps the entire interval between two adjacent maxima of the radius function to the enclosed minimum. For convenience we will assume that the interval is closed on the left and open on the right.

The push function gives us a succinct description of the task mechanics. In the next section we address the question of whether an arbitrary object can be oriented. Then we address the issue of how to model uncertainty, and after that we consider how to choose a sequence of pushes to orient an object.

Rotational symmetry. Orienting up to symmetry.

Consider again the example radius function of figure 7.22. The rotational symmetry of the rectangle results in a periodic radius function and a periodic push function. In such a case, there is no sequence of pushes that can unambiguously orient the object. The final

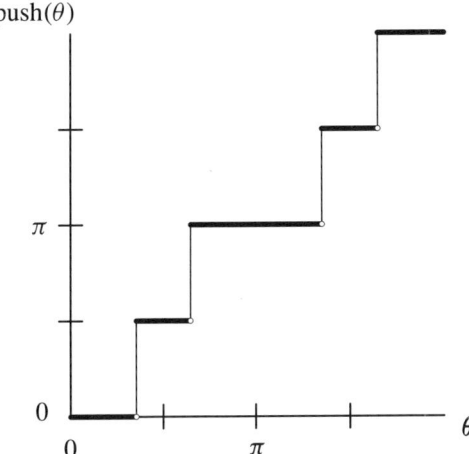

Figure 7.23
Example push function. Adapted from (Goldberg, 1993).

orientation will be determined only up to the underlying symmetry of the object. Note, however, that rotational symmetry of an object is in this case defined relative to the center of friction, so that an eccentric mass distribution can break the symmetry. Although we will not prove it here, Goldberg (1993) shows that every polygon can be oriented up to symmetry in its push function.

Modeling uncertainty

When the robot carries out a quasistatic plan, the task state is just the system configuration. Every execution of the plan can be modeled as a single trajectory through configuration space.

However, the parts orienting task necessarily involves uncertainty. If we imagine running the plan many times, and collect task states for all these runs in one plot, then we will see an "uncertainty cloud" of possible states. In devising the plan we have to think about these variations in state, and how to manipulate the entire uncertainty cloud.

The first question is, how to model the uncertainty—how to represent the ensemble of possible task states. For the present case, a simple approach is to consider closed intervals of the circle. A closed interval $\Theta = [\alpha, \beta]$ means that the actual state θ is a member of the interval. We also define a variant on the push function that operates on intervals: $\bar{p}(\Theta)$ returns the smallest interval containing $\{s(\theta) | \theta \in \Theta\}$.

This approach is sometimes called a *possibilistic* model. We have chosen to describe a set of possible states, without assigning probabilities to the states. In some cases a *probabilistic* model is preferable. If a possibilistic model is chosen, the problem now is how to represent configurations, and sets of configurations. Even in our simple case, we have chosen *not* to represent all possible subsets of the configuration space. Our restriction to closed intervals is an approximation that models any given set of possible states by a superset. It is readily observed that if we can find a plan that works for the approximating superset, it will also work for the true set of possible configurations, so we say that this is a *conservative* approximation.

Representing probability distributions in configuration space is generally more challenging, but in some instances there is good reason to do so.

Planning algorithm

Planning a sequence of pushes is straightforward, if you think backwards. Let ϕ_i be the pusher orientation for the ith step of the plan, and let Θ_i be the set of possible part orientations θ prior to the ith push. We want a sequence of n pushes, for some n, that orients the part up to symmetry. Equivalently, we will construct a sequence of pushes that maps from the largest possible interval Θ_1 to a single point.

We start with the *last* push, which maps from the interval Θ_n to a single orientation. What is the largest possible interval of orientations that is mapped by the push function to a single orientation? That is, what is the largest Θ such that $\bar{p}(\Theta)$ is a single point? By examining the push function in figure 7.23 we easily find the answer by looking for the largest step. That largest step, which we will call Θ_n, is a bound on the object orientations for which a single push will orient the object. (We don't know what n is yet, but we can still write Θ_n, Θ_{n-1}, etc.)

Now we repeat the process. We consider the next to last push, which must result in an object orientation within Θ_n. So now the question is, what is the largest interval Θ that is mapped to an interval smaller than Θ_n, i.e. $\bar{p}(\Theta) \subset \Theta_n$. That interval is the largest interval that can be unambiguously oriented in two pushes. Continuing in this way we obtain an algorithm:

1. Construct the push function.
2. Find the widest step in the push function. Call it Θ_n, and set i to 1.
3. Set Θ_{n-i} to the largest interval Θ such that $|\bar{p}(\Theta)| < |\Theta_{n-i+1}|$. If $|\Theta_{n-i}| = |\Theta_{n-i+1}|$ then set n to i and terminate. Otherwise increment i and repeat step 3.

Figures 7.24 and 7.25 illustrate the algorithm, using the rectangle of figures 7.22 and 7.23. First we find the widest step in the push function to obtain Θ_n, shown in

Quasistatic Manipulation

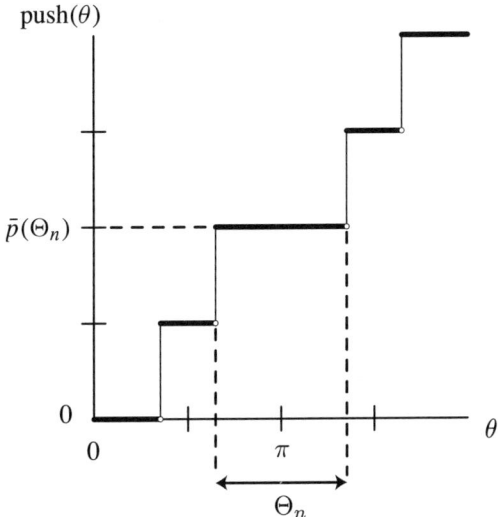

Figure 7.24
First step of the algorithm. We want the largest interval Θ such that $\bar{p}(\Theta)$ is a singleton. Adapted from (Goldberg, 1993).

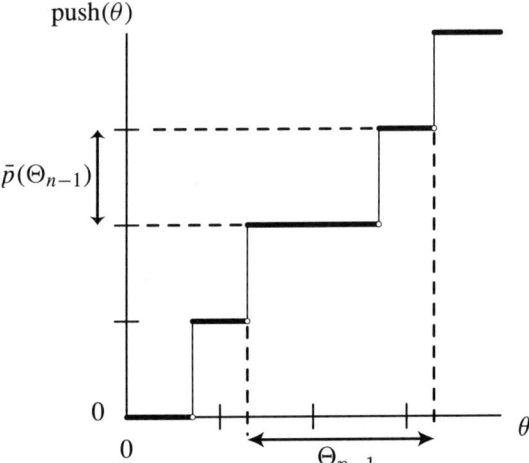

Figure 7.25
At subsequent steps of the algorithm, we look for the largest interval Θ such that $\bar{p}(\Theta)$ is smaller than the interval found in the previous step. Adapted from (Goldberg, 1993).

figure 7.24. Next we find the largest interval Θ that maps to a set $\bar{p}(\Theta)$ *smaller* than Θ_n. Figure 7.25 shows graphically the largest such interval, which is then Θ_{n-1}. In the construction of Θ_{n-2}, however, we find there is no interval larger than Θ_{n-1} mapping to a set smaller than Θ_{n-1}, so we terminate with $n = 2$. In this example we terminate with $|\Theta_{n-1}| = \pi$, due to symmetry in the push function. The rectangle will be oriented only up to a 180° symmetry. For generic objects the algorithm will terminate with Θ_1 equal to the entire circle.

The final step is to transform the plan into a useful form. The algorithm identifies the sequence of intervals Θ, which indirectly describes each push as range of orientations of the pushing support line relative to the object. A more useful description of the plan would be the sequence of supporting line orientations relative to some fixed frame.

Let ϕ_i be the pusher orientation relative to some fixed coordinate frame, for the ith step of the plan

1. Set ϕ_1 to zero; set i to 1.
2. Set ϕ_{i+1} to $push(\alpha_i) - \alpha_{i+1} - \epsilon_i + \phi_i$.

where $[\alpha_i, \beta_i] = \Theta_i$ and $\epsilon_i = \frac{1}{2}(|\Theta_i| - |\Theta_{i+1}|)$.

For the rectangle, ϕ_1 is zero, and ϕ_2 is $\frac{\pi}{4}$. A push, followed by a counterclockwise rotation of $\frac{\pi}{4}$ and another push, will orient the rectangle up to symmetry.

7.5 Assembly

Assembly is such a broad and complex problem that it encompasses all of the issues in manipulation. This section attempts to review briefly the broad topic of assembly, then focuses on quasistatic models of assembly. The assembly topic is taken up again in chapter 10, which addresses dynamic aspects of assembly tasks.

There are two ways of looking at assembly. First, we can look at assembly as an application task domain. As a central part of manufacturing automation, assembly is important not only because of the economic benefits, but also because it leads robotics research toward interesting problems.

Second, we can look at assembly as a fundamental process employed in many manipulation tasks. As we observed in chapter 1 the APOS system actually uses assembly to orient parts. Grasping a part is a kind of assembly. Even placing an object on a table is an assembly operation.

First we quickly review a few of the key issues in assembly:

- *Assembly sequence.* In what order should parts be joined to an assembly? In order to find a good sequence, we might first consider how to enumerate the different possible

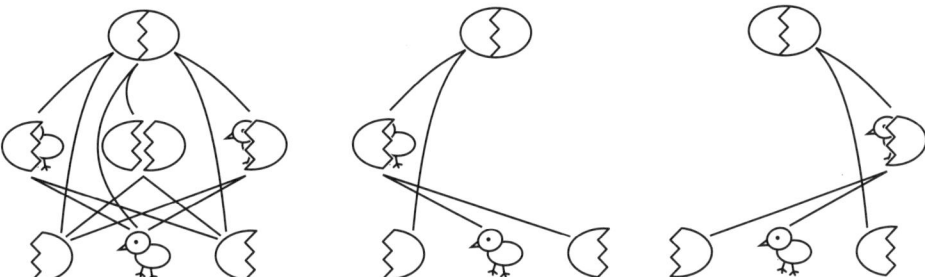

Figure 7.26
An and/or graph showing there are just two two-handed ways to unscramble an egg.

sequences. First we limit ourselves to *two-handed* assemblies, meaning that each assembly operation involves at most two independent motions, corresponding to two parts or subassemblies being brought together to form a new subassembly. To enumerate the possible assembly sequences, we look at every possible subassembly, and consider every possible way of partitioning it into two smaller subassemblies. The result can be represented as an *and/or* graph (figure 7.26). Each node of the graph is a subassembly. Each and-arc corresponds to a way of partitioning the subassembly. Any subtree with one and-arc per node constitutes a candidate assembly sequence. (Actually the subtree is a partial order, which still allows some freedom in the order of operations.)

- *Local constraint analysis.* Assembly by definition involves the process of bringing objects into kinematically constrained relationships. For any given assembly step, a key question is whether the final motion is consistent with all active kinematic constraints. For example, you cannot put the batteries in the flashlight after the lid is on. The easiest way to analyze this is to reverse time, and ask whether the disassembly is possible. A battery inside a closed flashlight is immobilized. Disassembly requires that the flashlight be opened before the batteries can be removed. How do we determine whether a disassembly motion exists? In many cases the first order form closure methods of previous chapters will suffice, although in some cases higher order analysis is necessary.

- *Path planning and grasp planning.* Where should the robot grasp parts and assemblies, and what path should it follow?

- *Gripper and fixture design.* Industrial assembly often involves specialized grippers designed to fit a particular feature of a particular part. In some instances, such as gripping a shaft, this is pure catalog work. In other cases standard designs will not suffice, and it is necessary to design a gripper that conforms to the particular part's shape. The same holds for fixture design. Both problems were addressed in section 7.1.

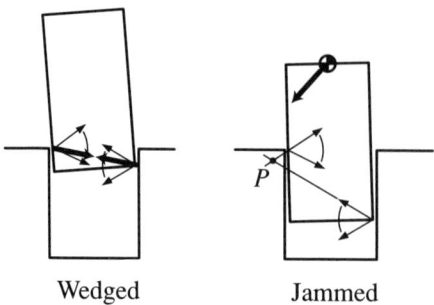

Figure 7.27
Wedging and jamming are two failure modes for assembly tasks.

- *Stable subassembly.* Given that we have placed two parts in the desired final relationship, will they stay in place while we move the other parts? As with path planning, this is a very hard problem to solve in the most general case. For simple cases the methods of form closure, force closure, and equilibrium analysis of previous chapters will suffice.

- *Assemblability.* Given that an assembly step is kinematically feasible, will the two parts go together without jamming? Jamming can prevent successful assembly, especially in the presence of shape tolerances, control error, and friction. Jamming is addressed below.

- *Tolerances.* Industrial parts are never perfect. Their shapes vary within limits. Also, there may be some variation possible in the relative position of two assembled parts. Tolerances are sometimes cumulative, so that "tolerance stackup" can lead to subassemblies with substantial variations in shape. In principle, all the analysis of local constraint, path planning, stability, and assemblability has to account for tolerances.

- *Design for assembly.* Our analytical tools are limited. The hardest assembly problems will be intractable for the foreseeable future. Fortunately, we can avoid the hardest problems by designing products and parts more intelligently, to simplify the analysis and planning of automated assembly problems. The APOS system discussed in chapter 1 was shaped in part by the design for assembly concept. Products can be designed so that almost all assembly steps are vertical insertions that can be performed by a manipulator with just four degrees of freedom. Parts can be designed so that orienting, grasping, and assembly steps all occur without jamming.

It should be clear that assembly is a big problem, and we can only touch on a few of the fundamental principles. The rest of this section addresses the quasistatic assembly of planar peg in hole, focusing on the mechanics of jamming and how to avoid jamming.

Quasistatic Manipulation

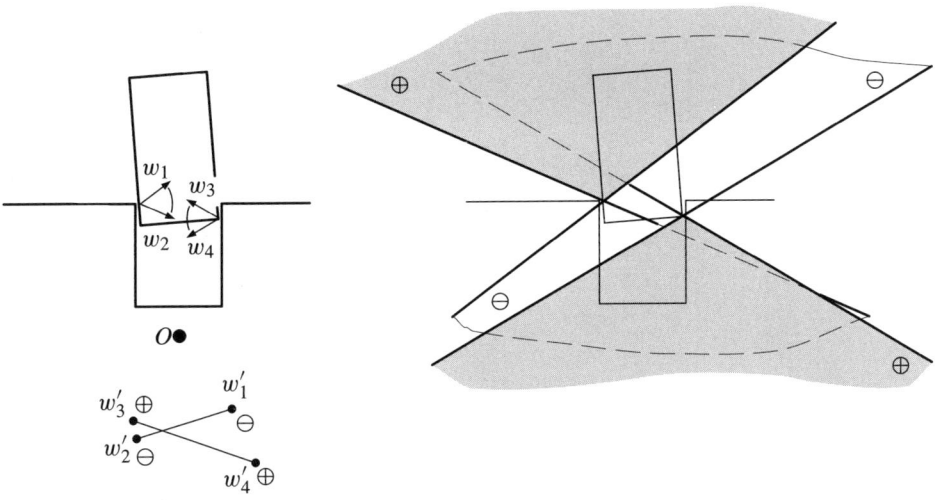

Figure 7.28
Force-dual and moment labeling analysis of the wedged peg. Force closure means there is no wrench that can be is guaranteed to move the peg under the assumptions of rigid bodies and Coulomb friction.

Jamming and wedging

This section summarizes research by Simunovic, Whitney, and colleagues. The first result by Simunovic (1975) analyzed the insertion of a planar peg in a planar hole. Simunovic identified two different ways that the peg could get stuck, which he named *wedging* and *jamming*. Figure 7.27 shows both cases. On the left is an example of wedging. Wedging is synonymous with force closure—no matter what wrench the robot applies to the peg, the frictional contacts between the peg and hole can balance it. Note that each friction cone contains the other's base, satisfying Nguyen's condition for force closure (see exercise 5.8). This means that the two contact forces can be arbitrarily large yet still be in balance. An additional insertion force can be balanced as well, by small perturbations of the contact forces. This is a phenomenon that everybody has experienced with drawers. A drawer that is pulled too far out, and then becomes cocked as it is pushed back in, can resist forces in any direction. Even attempts to pull the drawer back out may be fruitless.

The case on the right of figure 7.27 illustrates *jamming*. There is no force closure, but an unsuitable insertion wrench, such as that shown in the figure, will be balanced by the contact wrenches. Simunovic's analysis showed that if the applied line of action passes on the wrong side of the point P, it can be balanced by the contact forces to prevent assembly.

We can use the force dual and moment labeling methods to analyze the examples of figure 7.27. Figure 7.28 shows the wedged peg, the force dual construction, and the moment

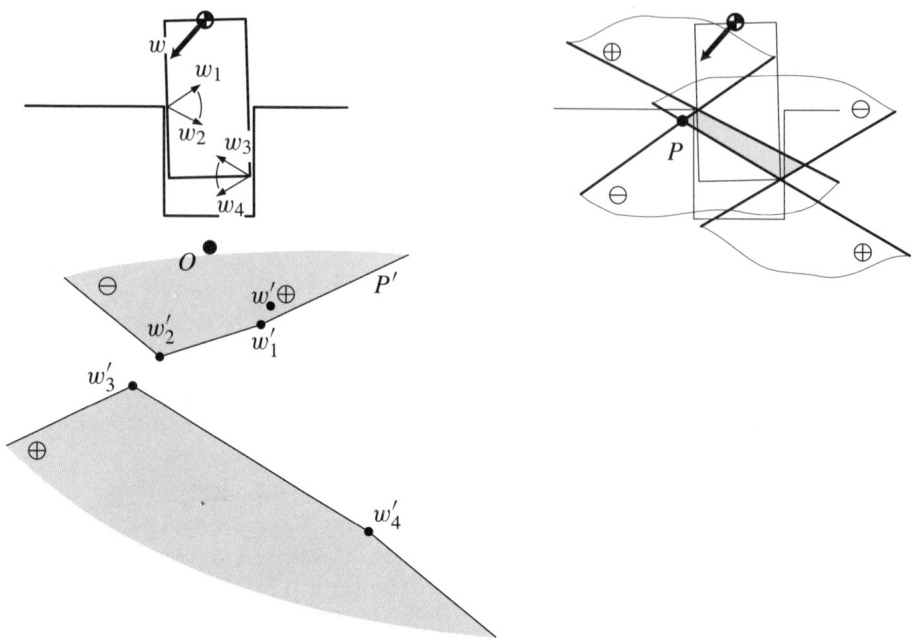

Figure 7.29
Force-dual and moment labeling analysis of the jammed peg. The applied wrench can be balanced by a resultant of the two frictional contacts.

labeling construction. This gives us two different ways of noticing that the peg is in force closure. The intersection of the moment labeling regions is null, meaning that the possible wrenches exhaust all of wrench space. And, the convex hull of the force dual regions is all of force dual space. Without a doubt, the peg is in force closure, and as long as we assume rigid bodies and Coulomb friction there is no applied wrench that will definitely move the peg.

Figure 7.29 shows the jammed peg, the force dual construction, and the moment labeling construction. Note that Simunovic's point P is one vertex of the moment labeling construction. In the force dual construction, we note that the applied wrench w maps into the region labeled with the opposite sign. Hence w can be balanced by the frictional contact forces. From the moment labeling construction we may draw the same conclusion.

Two lessons arise from the analysis of peg in hole. First, one can avoid wedging by avoiding two-point contact at shallow insertion depths. Beyond a certain depth Nguyen's condition cannot be satisfied. Second, to avoid jamming it is necessary to apply an insertion force that stays close to the peg.

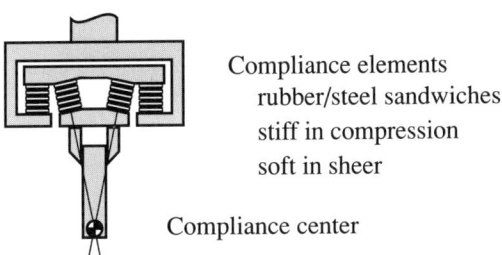

Compliance elements
rubber/steel sandwiches
stiff in compression
soft in sheer

Compliance center

Figure 7.30
A remote center of compliance device for industrial assembly tasks.

How do we arrange to have an appropriate insertion force—one that cannot be balanced by the frictional contacts between peg and hole? One approach is to use active force control of the robot, but in industrial applications a passive compliance is commonly used, and very neatly illustrates application of our analysis. We assume that the robot executes a programmed motion, with a passive compliance between the robot and the peg. We further assume the existence of a *compliance center*: the insertion force is assumed proportional to translational displacements of the center, and the torque about the compliance center is assumed proportional to angular displacements.

Where should we place the compliance center? In figure 7.29 the compliance center is at the top of the peg, which tends to occur naturally given the compliances contributed by the grasp and the structure of the robot itself. If we could place the compliance center near the mouth of the hole, or near the tip of the peg, the insertion force would pass through the labeled region of figure 7.29, and could not be balanced by the contact forces. Figure 7.30 shows a passive compliance called the Remote Center of Compliance or RCC, which projects a compliance center to the bottom of the peg, even though the peg is grasped at the top.

7.6 Bibliographic notes

The section on grasping and fixturing relies primarily on the work by Mishra, Schwartz, and Sharir (1987). A broader treatment would also address stability issues, compliance, non-point fingers, deformation of fingers and object, and measures of the quality of the grasp, to name just a few of the issues. Some of the key early work is that of Hanafusa and Asada (1977) and Asada and By (1985). Both grasp planning and fixture design continue to be active research areas. See (Bicchi and Kumar, 2000) for a survey of work on grasp and contact modeling. For recent work on fixture design see (Brost and Goldberg, 1996).

There is disagreement over the definitions of force closure and form closure. The terms originated in the kinematics of mechanisms, where it is necessary to distinguish joints that are held together by their form (such as the cylindrical pair of figure 2.21) or by some force (such as the planar pair of figure 2.21, assuming a gravitational field). This text chooses a different interpretation: "closure" implies complete restraint, and "force" versus "form" refers to the way we analyze a problem. These definitions are distilled from the most widely accepted usage in robotics research. Readers interested in a scholarly adventure should compare the definitions of Reuleaux (1876), Bicchi and Kumar (2000), Nguyen (1988), Lakshminarayana (1978), and Trinkle (1992).

The analysis of pushing presented in section 7.2 is my own work (1986). Pushing with a stable edge-edge contact is drawn from the work of Lynch and myself (1996), drawing heavily on the work of Peshkin and Sanderson (1988a). An alternative approach is to use the *minimum power principle* explored by Peshkin and Sanderson (1989) and Alexander and Maddocks (1993). From (Peshkin and Sanderson, 1989):

A quasistatic system chooses that motion, from among all motions satisfying the constraints, which minimizes the instantaneous power.

Some people find this principle so intuitively appealing that they are surprised to find it is not generally true. But, it is true for systems with no velocity dependent forces, such as the problems posed in this chapter. The minimum power principle makes an interesting comparison with the *maximum power inequality* (section 6.6). Nature maximizes power when choosing the frictional load for a given slider motion, but minimizes power when choosing the motion of a slider being pushed.

One practical problem is to determine the motion of pushed object when the support distribution is known. My PhD thesis (Mason and Salisbury, 1985) describes a numerical approach. Howe and Cutkosky (1996) address the problem using approximations of the limit surface. Alexander and Maddocks (1993) use the minimum power principle to solve the problem numerically, applying the methods of Overton (1983) to avoid convergence problems.

For the parts orienting method of section 7.4 the main source is (Goldberg, 1993). Earlier works on parts orienting include (Grossman and Blasgen, 1975), (Erdmann and Mason, 1988),(Brost, 1988),(Peshkin and Sanderson, 1988b). More recent work on parts orienting includes (Blind et al., 2000), which also includes a brief but useful survey.

The primary source for the mechanics of assembly is (Simunovic, 1975). The and/or graph and its use in enumerating assembly sequences is taken from (Homem de Mello and Sanderson, 1990). Two especially useful discussions of assembly are to be found in the works of (Latombe, 1991; Halperin et al., 1997). Also see De Fazio and Whitney's (1987) work for computer-aided assembly planning.

Quasistatic Manipulation

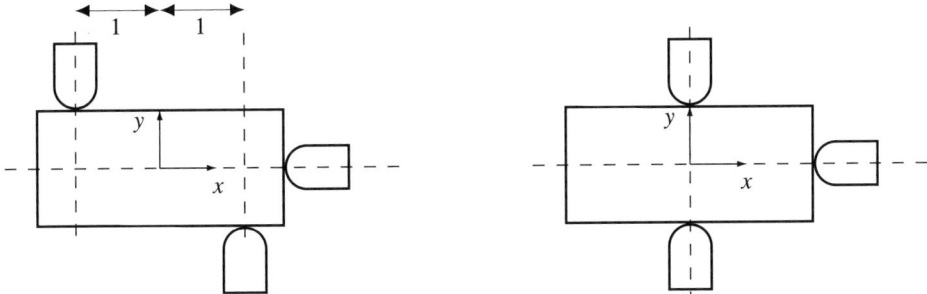

Figure 7.31
Two problems for grasp analysis (exercise 7.2).

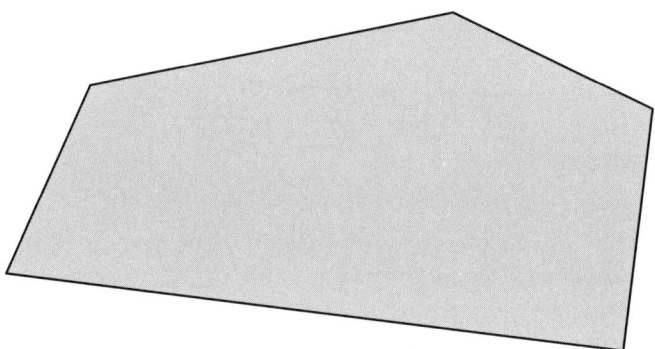

Figure 7.32
An object to be grasped with two fingers (exercise 7.3).

Exercises

Exercise 7.1: Suppose we lived in a four-dimensional Euclidean world. What is the smallest number of fingers that could achieve a first-order form closure grasp of a rigid object?

Exercise 7.2: Figure 7.31 shows two grasps of a rectangle in the plane by frictionless fingers.
(a) For each grasp draw the corresponding wrenches in wrench space.
(b) Are any of these grasps force closure? If not, show the location(s) of as many more fingers that it would take to achieve force closure. Use no more fingers than necessary.

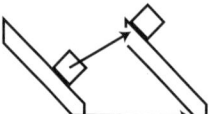

Figure 7.33
An apparatus for measuring the coefficient of friction by pushing.

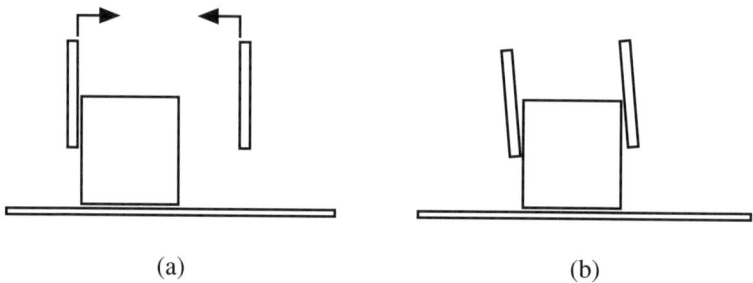

Figure 7.34
A pick operation to be analyzed (exercise 7.5).

Exercise 7.3: We want to find a force closure grasp for the object of figure 7.32 with just two fingers. If they were frictionless that would be impossible. In fact, there is some friction, but the coefficient is very small. Use the zigzag locus to find *all* two-finger grasps that give force closure with arbitrarily small but non-zero coefficient of friction.

Exercise 7.4: Figure 7.33 shows an apparatus for measuring the coefficient of friction. A pusher moves in a straight line, with a pushing surface sharply inclined from the direction of motion to ensure that slip occurs with the slider. Use a quasistatic force balance to show that the slider's motion relative to the contact normal gives the friction angle corresponding to the pusher-slider contact, regardless of the coefficient of friction between the slider and the support.

Exercise 7.5: Use the moment-labeling or the force-dual method to analyze the pick operation shown in figure 7.34.
  First, consider the situation shown in (a). Because the gripper is not perfectly centered, one finger makes contact first. We want the object to slide to a position centered between the fingers. Assuming the fingers can apply very large forces (but not infinite), determine a suitable range of coefficient of friction. You may assume the coefficient of friction is the same for all contacts.

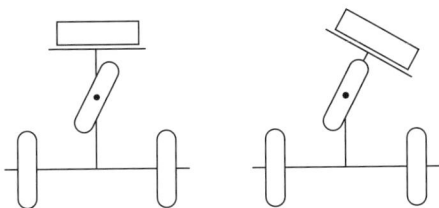

Figure 7.35
Two designs for a tricycle pusher (exercise 7.7).

Now, consider the situation shown in (b). Because the gripper is not perfectly aligned, we need the object to continue slipping so that it is aligned with the fingers. Again, find a suitable range for the coefficient of friction.

Finally, suppose the object is now aligned in the fingers, and has been lifted away from the table. We do not want the block to slip out of the fingers. Suppose the maximum finger force is actually 100 times the block weight. Find a suitable range for the coefficient of friction.

Exercise 7.6: The proof of the bisector bound that Randy Brost and I didn't publish is based on a result from (Mason and Salisbury, 1985). We consider the motion of a slider being pushed through a point contact, and we choose the origin to coincide with the point contact. The slider motion must produce zero frictional torque about the origin. We now consider the rotation center to be fixed, and define the function $g(\mathbf{r})$ to be the torque that would result from a unit normal force at $\mathbf{r}$. Then the total torque would be $\int g(\mathbf{r}) p(\mathbf{r}) \, dA = 0$. We plot the graph of g in x-y-g space. Let $g(R)$ be the locus of $(x, y, g(x, y))$ for all (x, y) in the support region R. Let G be the convex hull of $g(R)$. Let (x_0, y_0) be the center of friction. Then the given rotation center is a solution for some pressure distribution on R if and only if the point $(x_0, y_0, 0)$ is in G.

To prove the bisector bound, let the support region be the whole plane $R = \mathbf{E}^2$. Show that the resulting convex set G is strictly positive for all points (x, y) outside the circle centered on the rotation center and passing through the pushing contact. Show that the distance from the rotation center to the center of friction cannot exceed the distance from the rotation center to the pushing contact.

Exercise 7.7: Figure 7.35 shows two variations of a mobile robot designed for pushing. Each is a tricycle configuration with two unpowered rear wheels and one front wheel which is both steered and powered. On the left a snowplow blade is mounted to the frame of the tricycle, so that it is always parallel to the rear axle. On the right the snowplow blade is

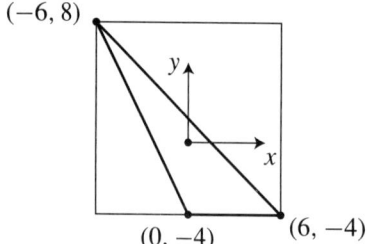

Figure 7.36
The Brost triangle.

mounted so that the blade is always parallel to the front axle. Suppose we are pushing the rectangle of figure 7.19, with the same coefficient of friction. For each tricycle, combine the stable pushing constraints of figure 7.19 with the constraints imposed by the tricycle, to identify the steering angles which produce stable pushing. Which design gives the greatest steering angle? How could that design be improved to give an even greater steering angle?

Exercise 7.8: The Brost triangle has vertices $(0, -4)$, $(6, -4)$, and $(-6, 8)$. Assuming a uniform mass distribution, the center of friction would then be at the origin. Suppose we want to push the Brost triangle with a stable edge-edge contact on the edge of intermediate length. Use the method described in the text to identify suitable instantaneous centers.

Exercise 7.9: Fill in the details of an algorithm to construct the radius function. Write and test code implementing the the algorithm. The input should be the coordinates for each vertex, and the angle of the supporting line. The output should be the directed distance to the supporting line.

Exercise 7.10: It was stated in section 7.4 that a minimum of the radius function can occur at the minimum of one of the individual sinusoids. This implies that the radius function can be negative, which for pushing problems would seem to require that the center of friction be *outside* the convex hull of the support region. For an example of where this might be useful, consider a mouse pushing a human-sized couch with four square legs. Although the center of friction would be inside the convex hull of the four legs, the mouse interacts with a single leg. Construct the radius function of a square, using a reference point that is on a diagonal, but outside the square.

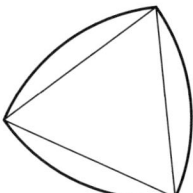

Figure 7.37
A figure of constant diameter—Reuleaux's triangle (exercise 7.12).

Exercise 7.11: The *diameter function* diam(θ) is related to the radius function. Given a bounded planar shape, the diameter at θ, diam(θ), is the distance between the two parallel supporting lines at angle θ. Show that the diameter function can be constructed as the maximum of a family of sinusoids. Construct the diameter function for the triangle of figure 7.36.

Exercise 7.12: Given an object's diameter function, can the object shape be reconstructed? (Hint: construct the diameter function of the circle, and of the object in figure 7.37. See (Reuleaux, 1876) for more examples of a similar nature.)

Exercise 7.13: Why are manhole covers round? (Hint: if you believed the answer you may have heard before—"It's the only shape that cannot fall into the manhole"—you should consider the previous exercise.)

Exercise 7.14: Consider the act of grasping a planar shape with a frictionless parallel jaw gripper. Let the *squeeze function* sqz(θ) be a function mapping the initial orientation to the final orientation of an object being squeezed in such a way. Assuming that the fingers make simultaneous contact, we can use the diameter function to construct the squeeze function, just as we used the radius function to construct the push function. Construct the squeeze function for the triangle of figure 7.36.

Exercise 7.15: Construct the radius function for a square with center of mass at one vertex. Can the object be oriented by pushes? Why or why not?

Exercise 7.16: Construct a planar polygonal shaped object, and choose a center of friction in the object's interior, so that the object is completely oriented with just one push. Construct the corresponding radius function.

Exercise 7.17: Construct the radius function for the triangle of figure 7.36 and show that maxima and minima are not isolated in this case. Suppose the goal is not just to orient the triangle, but to orient it so that it ends with the short side against the pusher. Find a coefficient of friction, and a sequence of pushes (not necessarily square pushes) to accomplish that goal.

Exercise 7.18: Construct the contact forces acting on the jammed peg of figure 7.29. Draw them the old fashioned way: as vectors at the contact points.

Exercise 7.19: Even though neither you nor your clothes are rigid bodies, it is still possible to use an and/or graph to enumerate potential procedures for getting dressed. For example, you can put your socks into your shoes before your feet, although it is difficult and the results are less than satisfactory. On the other hand, as far as I can determine it is impossible to put shorts on under long pants. Construct an and/or graph for getting dressed. Substitute generic symbols for the names of any unmentionable items, or leave them out entirely if you prefer. Research for this problem should not be done in class.

8 Dynamics

Dynamics plays two important roles in manipulation. It plays a direct role in the cases we termed *dynamic manipulation*: manipulation techniques that depend on inertial forces to accomplish a task. Dynamics also plays an indirect role in the case of *static manipulation* and *quasistatic manipulation*. Newtonian dynamics provides the foundation for static and quasistatic analysis. In particular, we appealed to the Newton-Euler equations in chapter 5 to provide a basis for the statics of rigid bodies in three-dimensional space.

This chapter reviews Newton's laws, develops the Newton-Euler equations for rigid bodies, and applies the theory to the motion of rigid bodies with frictional contact.

8.1 Newton's laws

In chapter 5 we hypothesized that a force can be described as a vector, and that the motion of a particle is determined by the vector sum of forces acting on the particle. Newton's laws, rendered in modern terms, provide additional hypotheses on the relation of force and motion:

1. Every body continues at rest, or in uniform motion in a straight line, unless forces act upon it.
2. The rate of change of momentum is proportional to the applied force.
3. The forces acting between two bodies are equal and opposite.

Newton's laws refer to *momentum*, which is for us a new term. To define momentum, consider a simple set of experiments, where different bodies are made to interact with some reference object. The nature of the interaction is not important, they might be connected by a spring, or by gravitation. For any given body, its acceleration will bear some fixed ratio to the acceleration of the reference object. We can take this ratio to be the *mass* of the object, and define its momentum to be the mass times the vector velocity.

8.2 A particle in three dimensions

First, we will consider a simple case, a particle in three dimensions. We choose some point in space to be the origin, and represent every point by a vector $\mathbf{x}$ from the origin to the point. So Newton's second law can be written

$$\frac{d\mathbf{p}}{dt} = \mathbf{F} \qquad (8.1)$$

where **p** is the momentum, and **F** is the total applied force. By definition, the momentum is the mass times the velocity,

$$\mathbf{p} = m\mathbf{v} = m\frac{d\mathbf{x}}{dt} \tag{8.2}$$

so for fixed mass the second law can also be written:

$$m\frac{d^2\mathbf{x}}{dt^2} = \mathbf{F}. \tag{8.3}$$

Integrating, we obtain a different form

$$\mathbf{p}_2 - \mathbf{p}_1 = \int_{t_1}^{t_2} \mathbf{F}\,dt \tag{8.4}$$

stating that the change in momentum is equal to the *impulse*.

We can also define *kinetic energy T*

$$T = \frac{m}{2}|\mathbf{v}|^2 \tag{8.5}$$

Differentiating yields

$$\frac{dT}{dt} = \frac{m}{2}\frac{d}{dt}(\mathbf{v}\cdot\mathbf{v}) \tag{8.6}$$

$$= \frac{m}{2}\left(\frac{d\mathbf{v}}{dt}\cdot\mathbf{v} + \mathbf{v}\cdot\frac{d\mathbf{v}}{dt}\right) \tag{8.7}$$

$$= m\frac{d\mathbf{v}}{dt}\cdot\mathbf{v} \tag{8.8}$$

$$= \mathbf{F}\cdot\mathbf{v} \tag{8.9}$$

stating that the time rate of change of kinetic energy is *power*. Integrating,

$$T_2 - T_1 = \int_{t_1}^{t_2} \mathbf{F}\cdot\mathbf{v}\,dt \tag{8.10}$$

or

$$T_2 - T_1 = \int_{\mathbf{x}_1}^{\mathbf{x}_2} \mathbf{F}\cdot d\mathbf{x} \tag{8.11}$$

stating that the change in kinetic energy is *work*.

Dynamics

8.3 Moment of force; moment of momentum

Recall that in chapter 5 we defined the moment of a force about a line, and about a point. The moment about the origin of a force **f** acting at a point **x** is given by

$$\mathbf{n} = \mathbf{x} \times \mathbf{f} \tag{8.12}$$

and the moment about a line l through the origin with direction $\hat{\mathbf{l}}$ is given by

$$n_l = \hat{\mathbf{l}} \cdot \mathbf{n} \tag{8.13}$$

Similarly, suppose a particle has momentum **p** passing through some point **x**. Then we can define the *moment of momentum*, or *angular momentum*, about the origin

$$\mathbf{L} = \mathbf{x} \times \mathbf{p} \tag{8.14}$$

and about some line l through the origin

$$L_l = \hat{\mathbf{l}} \cdot \mathbf{L} \tag{8.15}$$

(It seems that modern texts prefer the term "angular momentum", but "moment of momentum" is far more euphonius, and also reflects the commonalities with the moment of a directed line or the moment of force, arising from use of Plücker coordinates.)

Differentiating,

$$\frac{d\mathbf{L}}{dt} = \frac{d}{dt}(\mathbf{x} \times \mathbf{p}) \tag{8.16}$$

$$= \frac{d}{dt}(\mathbf{x} \times m\mathbf{v}) \tag{8.17}$$

$$= m\left(\frac{d\mathbf{x}}{dt} \times \mathbf{v} + \mathbf{x} \times \frac{d\mathbf{v}}{dt}\right) \tag{8.18}$$

$$= \mathbf{x} \times m\frac{d\mathbf{v}}{dt} \tag{8.19}$$

$$= \mathbf{x} \times \mathbf{F} \tag{8.20}$$

$$= \mathbf{N} \tag{8.21}$$

which is essentially a restatement of Newton's second law, but using moments of force and momentum. In integral form, we have

$$\mathbf{L}_2 - \mathbf{L}_1 = \int_{t_1}^{t_2} \mathbf{N}\,dt \tag{8.22}$$

Using either $\mathbf{F} = d\mathbf{p}/dt$ or $\mathbf{N} = d\mathbf{L}/dt$, we have three second order differential equations.

If **F** or **N** is uniquely determined by the state (**x**, **v**), then there is a unique solution giving **x**(*t*) and **v**(*t*) for any given initial conditions **x**(0) = **x**$_0$, **v**(0) = **v**$_0$.

8.4 Dynamics of a system of particles

Now we can extend Newton's laws to a system of particles. For the *k*th particle, define m_k to be the mass, $\mathbf{x}_k$ to be the position vector, and $\mathbf{p}_k$ to be the momentum. We divide the force $\mathbf{F}_k$ into two components: $\mathbf{F}_k^i$ the internal force, comprising the sum of all forces exerted by other particles in the system; and $\mathbf{F}_k^e$, the external force.

We define the momentum of the system to be

$$\mathbf{P} = \sum \mathbf{p}_k \tag{8.23}$$

and the total force on the system to be

$$\mathbf{F} = \sum \mathbf{F}_k^e \tag{8.24}$$

Note that the sum of all internal forces is zero, because by Newton's third law, the force exerted by one particle on another will be balanced by the reaction force.

For the *k*th particle Newton's second law yields

$$\frac{d\mathbf{p}_k}{dt} = \mathbf{F}_k^e + \mathbf{F}_k^i \tag{8.25}$$

Summing:

$$\sum \frac{d\mathbf{p}_k}{dt} = \sum \left(\mathbf{F}_k^e + \mathbf{F}_k^i \right) \tag{8.26}$$

Hence

$$\frac{d\mathbf{P}}{dt} = \mathbf{F} \tag{8.27}$$

showing that Newton's second law extends to the system of particles. If we define the total mass,

$$M = \sum m_k \tag{8.28}$$

and the center of mass,

$$\mathbf{X} = \frac{1}{M} \sum m_k \mathbf{x}_k \tag{8.29}$$

then we can write
$$\mathbf{P} = M \frac{d\mathbf{X}}{dt} \tag{8.30}$$
and
$$\mathbf{F} = M \frac{d^2\mathbf{X}}{dt^2} \tag{8.31}$$
which means that the center of mass behaves just like a single particle.

Moment of momentum and moment of force can also be extended to the system of particles. Define $\mathbf{L}_k$ to be the angular momentum of the kth point, define the total angular momentum to be the sum,
$$\mathbf{L} = \sum \mathbf{L}_k \tag{8.32}$$
and also define the total torque,
$$\mathbf{N} = \sum \mathbf{x}_k \times \mathbf{F}_k^e \tag{8.33}$$
Now for the kth particle, the angular momentum is given by
$$\mathbf{L}_k = m_k \mathbf{x}_k \times \mathbf{v}_k \tag{8.34}$$
Differentiating,
$$\frac{d\mathbf{L}_k}{dt} = m_k \mathbf{x}_k \times \frac{d\mathbf{v}_k}{dt} + m_k \frac{d\mathbf{x}_k}{dt} \times \mathbf{v}_k \tag{8.35}$$
The second term vanishes, leaving
$$\frac{d\mathbf{L}_k}{dt} = \mathbf{x}_k \times \frac{d\mathbf{p}_k}{dt} \tag{8.36}$$
Substituting Newton's second law for $\mathbf{p}_k$,
$$\frac{d\mathbf{L}_k}{dt} = \mathbf{x}_k \times \mathbf{F}_k^e + \mathbf{x}_k \times \mathbf{F}_k^i \tag{8.37}$$
Summing over all the particles,
$$\frac{d\mathbf{L}}{dt} = \mathbf{N} + \sum \mathbf{x}_k \times \mathbf{F}_k^i \tag{8.38}$$
Again, we can apply Newton's third law to show that the sum of the internal moments is zero, so that the second term vanishes:
$$\frac{d\mathbf{L}}{dt} = \mathbf{N} \tag{8.39}$$

showing that moment of force and moment of momentum behave just as they did for a single particle. For a system of particles, neither $\mathbf{F} = d\mathbf{P}/dt$, nor $\mathbf{N} = d\mathbf{L}/dt$, nor both sets of equations taken together, is sufficient to determine the motion of the particles. For that, we would have to take a detailed look at the particle interactions. However, the next section considers rigid bodies, where the particle interactions can be ignored.

8.5 Rigid body dynamics

As a final case, we will consider the dynamics of a system of particles, with the distances fixed—a rigid body. This restriction leads to motions of surprising elegance and complexity.

First, it is necessary to address a common misconception. Just as we often write Newton's second law as $\mathbf{F} = m \, d\mathbf{v}/dt$, it is tempting to write $\mathbf{N} = I \, d\boldsymbol{\omega}/dt$, where I is taken to be a scalar angular inertia, and $\boldsymbol{\omega}$ is taken to be a vector angular velocity. This is quite wrong, and reveals an important difference between linear motion and rotational motion under Newton's laws. The simplest case is with zero applied force and torque. Newton's first law says that, with zero applied force, the velocity of a body is constant. So one might speculate that, for zero applied torque, the angular velocity of a body is constant. This also is wrong. A body tumbling in space has a continually varying angular velocity, even though the angular momentum is constant. Any reader confronting this fact for the first time ought to try the experiment described in exercise 8.1.

We begin by deriving the correct equations of motion of a rotating rigid body. The velocity of a point x can be written

$$\mathbf{v} = \mathbf{v}_0 + \boldsymbol{\omega} \times \mathbf{x} \tag{8.40}$$

where $\mathbf{v}_0$ is the velocity of a point at the origin, and $\boldsymbol{\omega}$ is the angular velocity of the body.

Earlier, for the angular momentum of a point in a rotating body, we obtained equation 8.34:

$$\mathbf{L}_k = m_k \mathbf{x}_k \times \mathbf{v}_k$$

Substituting from equation 8.40:

$$\mathbf{L}_k = m_k \mathbf{x}_k \times (\mathbf{v}_0 + \boldsymbol{\omega} \times \mathbf{x}_k) \tag{8.41}$$

Summing to obtain the total angular momentum,

$$\mathbf{L} = \sum m_k \mathbf{x}_k \times \mathbf{v}_0 + \sum m_k \mathbf{x}_k \times (\boldsymbol{\omega} \times \mathbf{x}_k) \tag{8.42}$$

$$= M \mathbf{X} \times \mathbf{v}_0 + \sum m_k \mathbf{x}_k \times (\boldsymbol{\omega} \times \mathbf{x}_k) \tag{8.43}$$

The first term on the right hand side is the angular momentum that would be obtained if all the mass were concentrated at the center of mass. We can eliminate this term by placing the origin at the center of mass, yielding:

$$\mathbf{L} = \sum m_k \mathbf{x}_k \times (\boldsymbol{\omega} \times \mathbf{x}_k) \tag{8.44}$$

Applying the identity $\mathbf{a} \times (\mathbf{b} \times \mathbf{c}) = (\mathbf{a} \cdot \mathbf{c})\mathbf{b} - (\mathbf{a} \cdot \mathbf{b})\mathbf{c}$,

$$\mathbf{L} = \sum m_k \left[(\mathbf{x}_k \cdot \mathbf{x}_k)\boldsymbol{\omega} - \mathbf{x}_k (\mathbf{x}_k \cdot \boldsymbol{\omega}) \right] \tag{8.45}$$

To factor $\boldsymbol{\omega}$ out of the sum, we can represent each vector as a column matrix, and substitute $\mathbf{x}_k^t \boldsymbol{\omega}$ for $\mathbf{x}_k \cdot \boldsymbol{\omega}$:

$$\mathbf{L} = \left(\sum m_k \left(|\mathbf{x}_k|^2 I_3 - \mathbf{x}_k \mathbf{x}_k^t \right) \right) \boldsymbol{\omega} \tag{8.46}$$

where I_3 is the three-by-three identity matrix. We now define the *angular inertia matrix* I:

$$I = \sum m_k \left(|\mathbf{x}_k|^2 I_3 - \mathbf{x}_k \mathbf{x}_k^t \right) \tag{8.47}$$

Substituting above,

$$\mathbf{L} = I \boldsymbol{\omega} \tag{8.48}$$

Substituting equation 8.48 into $\mathbf{N} = d\mathbf{L}/dt$ yields

$$\mathbf{N} = \frac{dI\boldsymbol{\omega}}{dt} \tag{8.49}$$

$$= I \frac{d\boldsymbol{\omega}}{dt} + \frac{dI}{dt} \boldsymbol{\omega} \tag{8.50}$$

From inspection of equation 8.47, it is apparent that the inertia matrix is constant with respect to a coordinate frame fixed to the body. But the body is rotating with angular velocity $\boldsymbol{\omega}$, so the inertia matrix is time-varying in any fixed frame. Each column vector in the inertia matrix can be treated as a vector fixed in the moving body, so that the velocity of each column vector is given by taking cross product with $\boldsymbol{\omega}$, which can also be expressed as matrix product with the cross-product matrix W:

$$W = \begin{pmatrix} 0 & -\omega_3 & \omega_2 \\ \omega_3 & 0 & -\omega_1 \\ -\omega_2 & \omega_1 & 0 \end{pmatrix} \tag{8.51}$$

Hence

$$\mathbf{N} = I\frac{d\boldsymbol{\omega}}{dt} + WI\boldsymbol{\omega} \tag{8.52}$$

$$\mathbf{N} = I\frac{d\boldsymbol{\omega}}{dt} + W(I\boldsymbol{\omega}) \tag{8.53}$$

$$\mathbf{N} = I\frac{d\boldsymbol{\omega}}{dt} + \boldsymbol{\omega} \times (I\boldsymbol{\omega}) \tag{8.54}$$

Equation 8.54 is Newton's second law applied to rigid body rotation. Note that the analogy with linear motion fails in two ways: the angular inertia is a matrix rather than a scalar; and there is a second term on the right hand side, $\boldsymbol{\omega} \times (I\boldsymbol{\omega})$. The presence of this term causes a time-varying angular velocity, even when $\mathbf{N} = \mathbf{0}$.

Equation 8.54 has an elegant form when expanded in components, provided that we choose the right coordinate frame. In the next section we will see that, if we choose as our coordinate frame the *principal axes* of the body, then the angular inertia matrix is diagonal:

$$I = \begin{pmatrix} I_{11} & 0 & 0 \\ 0 & I_{22} & 0 \\ 0 & 0 & I_{33} \end{pmatrix} \tag{8.55}$$

We want to use this diagonal form, but we need a fixed coordinate frame. Define $\{I_t\}$ to be the fixed coordinate frame that coincides with the principal body coordinates at time t. Then at any time t, in coordinate frame $\{I_t\}$, we can expand equation 8.54 to obtain:

$$\begin{pmatrix} N_1 \\ N_2 \\ N_3 \end{pmatrix} = \begin{pmatrix} I_{11} & 0 & 0 \\ 0 & I_{22} & 0 \\ 0 & 0 & I_{33} \end{pmatrix} \begin{pmatrix} \dot{\omega}_1 \\ \dot{\omega}_2 \\ \dot{\omega}_3 \end{pmatrix}$$

$$+ \begin{pmatrix} 0 & -\omega_3 & \omega_2 \\ \omega_3 & 0 & -\omega_1 \\ -\omega_2 & \omega_1 & 0 \end{pmatrix} \begin{pmatrix} I_{11} & 0 & 0 \\ 0 & I_{22} & 0 \\ 0 & 0 & I_{33} \end{pmatrix} \begin{pmatrix} \omega_1 \\ \omega_2 \\ \omega_3 \end{pmatrix}$$

which can be expanded to give *Euler's equations*:

$$\begin{pmatrix} N_1 \\ N_2 \\ N_3 \end{pmatrix} = \begin{pmatrix} I_{11}\dot{\omega}_1 + (I_{33} - I_{22})\omega_2\omega_3 \\ I_{22}\dot{\omega}_2 + (I_{11} - I_{33})\omega_3\omega_1 \\ I_{33}\dot{\omega}_3 + (I_{22} - I_{11})\omega_1\omega_2 \end{pmatrix} \tag{8.56}$$

We derived Euler's equations assuming a fixed coordinate frame, but they also hold in the moving principal axis coordinate system. To see that this is so, suppose we have two observers placed at the center of mass. One is fixed in the body, the other is motionless. At some certain time t, they happen to coincide instantaneously. Now imagine that the

Dynamics

two observers have two different geometrical vectors in view: the torque vector **N** and the angular velocity vector ω. Since the observers are coincident, both of them report identical coordinates for the two vectors. But suppose we ask them about the rate of change of a vector. Since the two observers are in relative motion, they will generally give different answers. For example, a vector fixed with respect to the fixed observer will generally appear to be moving with respect to the moving observer. But this is *not* the case for the angular velocity vector ω, because the observer's rotation axis is that same vector. The rotating observer and the fixed observer see the same rate of change in this case. So Euler's equations describe the angular velocity ω with respect to the moving principal axes coordinate frame.

We can also find a simple expression for the kinetic energy T in a rotating rigid body. The body kinetic energy is the sum of the kinetic energy at the point masses:

$$T = \sum \frac{1}{2} m_k |\mathbf{v}_k|^2 \tag{8.57}$$

$$= \sum \frac{1}{2} m_k \mathbf{v}_k \cdot \mathbf{v}_k \tag{8.58}$$

$$= \sum \frac{1}{2} m_k (\omega \times \mathbf{x}_k) \cdot (\omega \times \mathbf{x}_k) \tag{8.59}$$

where we are assuming that the velocity at the origin is zero. Now, using the triple product identity $a \cdot b \times c = b \cdot c \times a$, we obtain

$$T = \sum \frac{1}{2} m_k \omega \cdot (\mathbf{x}_k \times (\omega \times \mathbf{x}_k)) \tag{8.60}$$

$$= \frac{1}{2} \omega \cdot \mathbf{L} \tag{8.61}$$

$$= \frac{1}{2} \omega \cdot I \omega \tag{8.62}$$

8.6 The angular inertia matrix

This section describes techniques for simplifying the description and construction of angular inertia matrices. Generalizing slightly from equation 8.47, we will consider the angular inertia matrix for a continuous body:

$$I = \int \rho \left(|\mathbf{x}|^2 I_3 - \mathbf{x}\mathbf{x}^t \right) dV \tag{8.63}$$

where ρ is the density of the material, and dV is a differential element of volume.

Some insights can be gained by expanding the inertia matrix into components:

$$I = \int \rho \begin{pmatrix} x_2^2 + x_3^2 & -x_1 x_2 & -x_1 x_3 \\ -x_1 x_2 & x_1^2 + x_3^2 & -x_2 x_3 \\ -x_1 x_3 & -x_2 x_3 & x_1^2 + x_2^2 \end{pmatrix} dV \qquad (8.64)$$

The diagonal elements are the moments of inertia with respect to the three coordinate axes:

$$I_{11} = \int \rho (x_2^2 + x_3^2) \, dV \qquad (8.65)$$

$$I_{22} = \int \rho (x_1^2 + x_3^2) \, dV \qquad (8.66)$$

$$I_{33} = \int \rho (x_1^2 + x_2^2) \, dV \qquad (8.67)$$

The off-diagonal terms are called the *products of inertia*:

$$I_{12} = I_{21} = -\int \rho x_1 x_2 \, dV \qquad (8.68)$$

$$I_{23} = I_{32} = -\int \rho x_2 x_3 \, dV \qquad (8.69)$$

$$I_{31} = I_{13} = -\int \rho x_3 x_1 \, dV \qquad (8.70)$$

It is readily apparent that the inertia matrix is symmetric, which makes further simplification possible. There is some choice of coordinate frame for which the inertia matrix is a diagonal matrix. That is, there is some coordinate frame A, which gives a diagonal inertia matrix:

$$^{A}I = \begin{pmatrix} ^{A}I_{11} & 0 & 0 \\ 0 & ^{A}I_{22} & 0 \\ 0 & 0 & ^{A}I_{33} \end{pmatrix} \qquad (8.71)$$

The description in A-coordinates can be obtained by the transformation:

$$^{A}I = A I A^{T} \qquad (8.72)$$

where A is the coordinate transform matrix from base coordinates to A-coordinates. If the angular inertia matrix is taken with respect to an origin at the body center of mass, the frame A that diagonalizes the inertia matrix has a special significance. The coordinate axes of this frame, which are the eigenvectors of the matrix, are called the *principal axes of the body*. The I_{11}, I_{22}, and I_{33} elements of the diagonal matrix, which are the eigenvalues of the matrix, are called the *principal moments of inertia*. When these eigenvalues are unique,

Dynamics

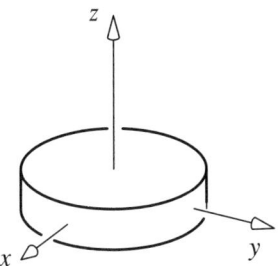
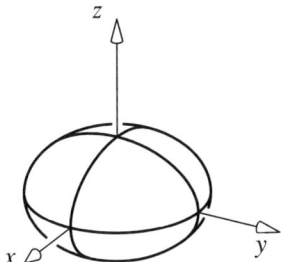

Figure 8.1
A cylinder and its inertia ellipsoid.

the corresponding eigenvectors are uniquely determined. When the eigenvalues are not unique, there is some freedom in the choice of principal axes.

In some cases we are not concerned with the entire inertia matrix, but just with a scalar angular inertia about a certain direction. For example, for an object that is mounted so that rotation is only possible about a fixed axis with direction $\hat{\mathbf{n}}$, the rotational mechanics is described by the equation

$$N_n = I_n \ddot{\theta}_n \qquad (8.73)$$

where N_n is the moment of force about the fixed axis, I_n is the angular inertia about the fixed axis, and θ_n is the angle about the fixed axis. The scalar angular inertia about the axis with direction $\hat{\mathbf{n}}$ is given by

$$I_n = \hat{\mathbf{n}}^t I \hat{\mathbf{n}} \qquad (8.74)$$

We can also define a *radius of gyration* k_n with respect to the axis:

$$I_n = M k_n^2 \qquad (8.75)$$

The radius of gyration represents the distance of a point mass that would give the same angular inertia.

Finally, there is a geometrical characterization of the inertia matrix. Consider the surface described by the equation

$$\mathbf{r}^t I \mathbf{r} = a \qquad (8.76)$$

where a is some arbitrary constant, typically 1. In principal axis coordinates, this becomes

$$I_{xx} r_x^2 + I_{yy} r_y^2 + I_{zz} r_z^2 = a \qquad (8.77)$$

Since I_{xx}, I_{yy}, and I_{zz} are all greater than zero, except for degeneracies, this gives the equation of an ellipsoid, called the *inertia ellipsoid* (figure 8.1). Let $\mathbf{r} = r\hat{\mathbf{n}}$ be some vector from the center of mass to the surface of the inertia ellipsoid. Then we have

$$I_n = \hat{\mathbf{n}}^t I \hat{\mathbf{n}} = \frac{1}{r^2}\mathbf{r}^t I \mathbf{r} = \frac{a}{r^2} \tag{8.78}$$

Comparing with equation 8.75,

$$Mk_n^2 = \frac{a}{r^2} \tag{8.79}$$

So in any direction, the distance to the surface of the ellipsoid is inversely related to the radius of gyration about the same axis.

Given some body, how do we construct the inertia matrix? The most direct method is to place the origin at the center of mass, then evaluate the six integrals in equation 8.64. But there are considerably simpler ways. Most of the problems that one finds in textbooks involve symmetric objects. By identifying these symmetries, you can choose a coordinate frame that gives a diagonal matrix, reducing the integrals from six to three.

THEOREM 8.1:   Any plane of symmetry is perpendicular to a principal axis.

Proof   Exercise 8.3.   ∎

THEOREM 8.2:   Any line of symmetry is a principal axis. The other two principal axes can be chosen arbitrarily, provided the three are orthogonal.

Proof   Exercise 8.3.   ∎

Of course, many objects do not have a plane of symmetry or a line of symmetry. However, some complex objects can be decomposed into simpler parts that do exhibit symmetry. The inertia matrix of the composite object is just the sum of the inertia matrices of the parts, but they must all be expressed in the same coordinate system. So the procedure is to find each part's inertia matrix as if were an isolated body, transform all these inertia matrices to a common coordinate frame at the composite center of mass, and add them up. The transformation of an inertia matrix to the common coordinate frame may require a rotation (equation 8.72) and also a change of origin, requiring one more theorem.

Dynamics

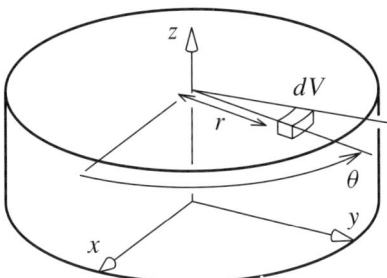

Figure 8.2
Example: calculating the intertia matrix for a cylinder.

THEOREM 8.3 PARALLEL AXIS THEOREM: Let I be the inertia matrix of a body at its center of mass 0, and let ^pI be the inertia matrix at some point p. Then

$$^pI = I + M(|\mathbf{p}|^2 I_3 - \mathbf{p}\mathbf{p}^t) \tag{8.80}$$

where M is the total mass of the body.

Proof   Exercise 8.4. ∎

Note the similarity of equation 8.80 to the definition of the inertia matrix in equation 8.47. The inertia matrix at p is the sum of the original matrix, and a second matrix describing a point mass at the body center of mass. Returning to the problem of a composite body, the total inertia matrix can be found as the sum of $n + 1$ matrices—1 matrix for each of the n parts, and an additional matrix describing the inertia of a body obtained by substituting a point mass for each of the parts.

EXAMPLE
Find the angular inertia matrix for a cylinder of uniform density $\rho = 1$, height 1, and radius 1.

In this example, both symmetry theorems apply. The axis of the cylinder is an axis of symmetry, and the perpendicular plane through the center of mass is a plane of symmetry. One principal axis lies along the cylinder axis, and the other two lie in the perpendicular plane. If the cylinder axis is taken as the $\hat{\mathbf{z}}$-axis, then due to the symmetry, I_{xx} and I_{yy} are equal. We will start with I_{xx}, which we decompose into two integrals:

$$I_{xx} = \int \rho(y^2 + z^2)\, dV \tag{8.81}$$

$$= \int \rho y^2\, dV + \int \rho z^2\, dV \tag{8.82}$$

This integral is equivalent to an iterated integral using cylindrical coordinates (r, θ, z) (figure 8.2).

$$x = r\cos\theta \tag{8.83}$$
$$y = r\sin\theta \tag{8.84}$$
$$z = z \tag{8.85}$$

and with the differential element of volume given by

$$dV = r\, dr\, d\theta\, dz \tag{8.86}$$

Then we obtain

$$\int \rho y^2\, dV = \int_{-1/2}^{1/2} \int_0^{2\pi} \int_0^1 \rho r^3 \sin^2\theta\, dr\, d\theta\, dz \tag{8.87}$$

$$= \rho\pi/4 \tag{8.88}$$

and

$$\int \rho z^2\, dV = \int_{-1/2}^{1/2} \int_0^{2\pi} \int_0^1 \rho z^2 r\, dr\, d\theta\, dz \tag{8.89}$$

$$= \rho\pi/12 \tag{8.90}$$

so that

$$I_{xx} = \rho\pi/4 + \rho\pi/12 = \pi/3 \tag{8.91}$$

Similarly we can find I_{zz} by decomposing the integral

$$I_{zz} = \int \rho x^2\, dV + \int \rho y^2\, dV \tag{8.92}$$

but from symmetry, $\int \rho y^2 \, dV$ is equal to $\int \rho x^2 \, dV$, which was already determined to be $\pi/4$. So the principal moments of inertia are

$$I_{xx} = \pi/3 \tag{8.93}$$
$$I_{yy} = \pi/3 \tag{8.94}$$
$$I_{zz} = \pi/2 \tag{8.95}$$

8.7 Motion of a freely rotating body

If the inertia matrix I is not degenerate, and if the force **F** and torque **N** are well behaved, then the Newton-Euler equations

$$\mathbf{F} = m \frac{d\mathbf{v}}{dt} \tag{8.96}$$
$$\mathbf{N} = I \frac{d\boldsymbol{\omega}}{dt} + \boldsymbol{\omega} \times (I \boldsymbol{\omega}) \tag{8.97}$$

will uniquely determine the motion of a rotating body. In this section we consider the case of a body with zero applied force or torque

$$\mathbf{F} = \mathbf{0} \tag{8.98}$$
$$\mathbf{N} = \mathbf{0} \tag{8.99}$$

We already know that velocity of the center of mass will be constant—fixed if we choose the right frame. But how can we describe the rotational motion?

We can set **N** equal to zero in Euler's equations (8.56), obtaining

$$\dot{\omega}_1 = \frac{I_2 - I_3}{I_1} \omega_2 \omega_3 \tag{8.100}$$
$$\dot{\omega}_2 = \frac{I_3 - I_1}{I_2} \omega_3 \omega_1 \tag{8.101}$$
$$\dot{\omega}_3 = \frac{I_1 - I_2}{I_3} \omega_1 \omega_2 \tag{8.102}$$

We can readily identify cases that will leave the angular velocity constant. For instance, if $I_1 = I_2 = I_3$, then there will be no angular acceleration. Similarly, if two principal moments, say I_1 and I_2, and the corresponding angular velocity, ω_3 in this case, were initially zero, then there would be no angular acceleration. The third case is where two components of the angular velocity are zero: If the angular velocity is parallel to a principal axis, there will be no angular acceleration. (However, the equilibrium might be unstable. See exercise 8.2.)

In general, it should now be obvious, the angular velocity will not be constant. But, with zero applied torque, the angular momentum **L** is constant. Similarly, with zero applied torque, the kinetic energy is constant. For **N** = **0**, then:

$$\mathbf{L} = I\boldsymbol{\omega} \quad \text{is constant} \tag{8.103}$$

$$T = \frac{1}{2}\boldsymbol{\omega}^t I \boldsymbol{\omega} \quad \text{is constant} \tag{8.104}$$

Equation 8.104 is the equation giving the surface of the inertia ellipsoid. As the body tumbles, the inertia ellipsoid tumbles as well, and the angular velocity vector is constantly varying, but it always falls on the surface of the inertia ellipsoid.

We can determine the plane tangent to the inertia ellipsoid at $\boldsymbol{\omega}$. First, we find the direction of the normal to the plane, by taking the gradient of T. Using principal axis coordinates:

$$\nabla \frac{1}{2} \boldsymbol{\omega}^t I \boldsymbol{\omega} = \nabla \frac{1}{2} (\omega_1^2 I_1 + \omega_2^2 I_2 + \omega_3^2 I_3) \tag{8.105}$$

$$= (I_1 \omega_1, I_2 \omega_2, I_3 \omega_3) \tag{8.106}$$

$$= \mathbf{L} \tag{8.107}$$

So the angular momentum **L** is perpendicular to the plane tangent to the inertia ellipsoid at $\boldsymbol{\omega}$. Next, we find the distance from the center of mass to the tangent plane (figure 8.3):

$$\frac{\boldsymbol{\omega} \cdot \mathbf{L}}{|\mathbf{L}|} = \frac{2T}{|\mathbf{L}|} \tag{8.108}$$

which is also constant. The plane's orientation doesn't change, and its distance from the origin does not change, which means that the plane does not vary during the motion. It is called the *invariable plane*.

This leads to an elegant geometrical description of the motion of a tumbling body, known as *Poinsot's construction* (figure 8.3). Imagine that the inertia ellipsoid is a solid body, whose center is fixed, and which is in contact with a solid invariable plane. Because the angular velocity vector passes through the contact point, that point is instantaneously motionless. Thus the motion of the body is equivalent to the rolling of the inertia ellipsoid, without slip, on the invariable plane. The path of the contact point on the inertia ellipsoid is called the *polhode*, and the curve traced on the invariable plane is called the *herpolhode*. Note that if we project rays from the center of mass through the polhode, and the herpolhode, we obtain the moving cone, and the fixed cone, respectively, of the rotation

Dynamics

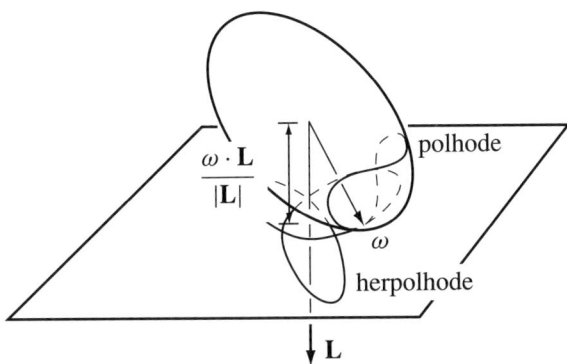

Figure 8.3
Poinsot's construction: the inertia ellipsoid rolls without sliding on the invariable plane. The contact points trace out two curves: the polhode and herpolhode.

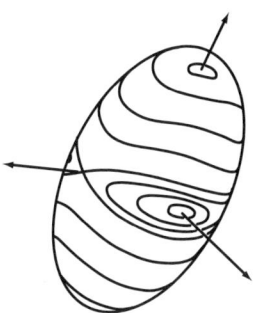

Figure 8.4
The polhodes on a generic inertia ellipsoid. Note the difference between the intermediate axis, where the polhodes diverge, and the other axes, where the polhodes stay close.

(section 2.3). For symmetric objects, the polhodes are circles, and the cones are true circular cones.

8.8 Planar single contact problems

Consider again the mechanics of a pair of particles, subject to Newton's laws with, say, elastic forces acting between. The state of the system is given by the positions and velocities of the particles. The forces are completely determined by the state of the system. The future of the system can be predicted by integrating the forces. Generally speaking,

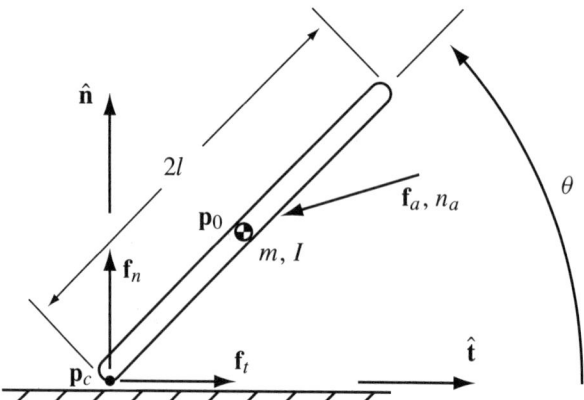

Figure 8.5
Notation for the sliding rod problem.

when force is a well behaved function of position and velocity, Newton's laws will have a unique solution.

Now suppose that we add some frictional forces, governed by Coulomb's law. The table of contact modes in section 6.3 does not give frictional forces as a function of state. Rather, it imposes *constraints* on the forces, as a function of position, velocity, *and acceleration*. We have a circularity: accelerations are determined by forces, which are now determined partly by accelerations. A systematic approach to friction problems has to face this circularity squarely.

The basic approach we will take is quite simple, in principle:

1. Enumerate the contact modes.

2. Solve the mechanics problem associated with each contact mode to obtain velocities and accelerations at each contact point.

3. Discard the solution if the contact velocities and accelerations are not consistent with the hypothesized contact mode.

This is certainly a strange way to approach a mechanics problem. We break it into several problems, and solve them all, even though ultimately only one turns out to be relevant. Now we get to something even stranger: sometimes more than one of the subproblems works; and sometimes none of the subproblems work. Newtonian mechanics with rigid bodies and Coulomb friction is sometimes *nondeterministic* and sometimes *inconsistent*.

Besides the philosophical difficulties with our approach, there is also a practical difficulty: there can be lots of contact modes, so solving a separate problem for each can be lots of work. Is it possible there might be a more efficient way of identifying the relevant contact modes? The problem is more difficult when there are several moving bodies, because the number of contact modes grows exponentially with the number of bodies. Baraff (1990, 1993) has shown that determining whether any consistent contact mode exists is NP-complete in the case of multiple bodies with Coulomb friction. This means that the problem probably cannot be solved efficiently, although we might find an approximate solution or an alternative set of assumptions that would be satisfactory.

Example

To illustrate the approach, we will explore the mechanics of a rod with a single contact on a horizontal surface (figure 8.5). We can save some work by writing the equations of motion in terms of the contact point $\mathbf{p}_c$. First, we note some kinematic relations:

$$\mathbf{p}_c = \mathbf{p}_0 - l \begin{pmatrix} \cos\theta \\ \sin\theta \end{pmatrix} \tag{8.109}$$

$$\dot{\mathbf{p}}_c = \dot{\mathbf{p}}_0 - l \begin{pmatrix} -\sin\theta \\ \cos\theta \end{pmatrix} \tag{8.110}$$

$$\ddot{\mathbf{p}}_c = \ddot{\mathbf{p}}_0 + l \begin{pmatrix} \cos\theta \\ \sin\theta \end{pmatrix} \dot{\theta}^2 - l \begin{pmatrix} -\sin\theta \\ \cos\theta \end{pmatrix} \ddot{\theta} \tag{8.111}$$

Now, applying Newton's laws:

$$\mathbf{f}_n + \mathbf{f}_t + \mathbf{f}_a = m\ddot{\mathbf{p}}_0 \tag{8.112}$$

$$(\mathbf{p}_c - \mathbf{p}_0) \times (\mathbf{f}_n + \mathbf{f}_t) + n_a = I\ddot{\theta} \tag{8.113}$$

Solving for $\ddot{\mathbf{p}}_0$ and $\ddot{\theta}$ and substituting into equation 8.111:

$$\ddot{\mathbf{p}}_c = \frac{1}{m}(\mathbf{f}_n + \mathbf{f}_t + \mathbf{f}_a) + l\dot{\theta}^2 \begin{pmatrix} \cos\theta \\ \sin\theta \end{pmatrix}$$

$$- \frac{l}{I} \begin{pmatrix} -\sin\theta \\ \cos\theta \end{pmatrix} [(\mathbf{p}_c - \mathbf{p}_0) \times (\mathbf{f}_n + \mathbf{f}_t) + n_a] \tag{8.114}$$

Equation 8.114 is a general equation of motion. Before we can solve it, we need values for some physical parameters, and we need initial conditions. In the next two sections we show parameter choices to get an inconsistency (no solutions) and parameter choices yielding an ambiguity (multiple solutions).

Frictional inconsistency

Suppose we have a gravitational field, and the rod is initially sliding toward the left without rotating. Then we have

$$\dot{\mathbf{p}}_0 = \begin{pmatrix} -1 \\ 0 \end{pmatrix} \tag{8.115}$$

$$\dot{\theta} = 0 \tag{8.116}$$

$$\mathbf{f}_a = \begin{pmatrix} 0 \\ -mg \end{pmatrix} \tag{8.117}$$

$$n_a = 0 \tag{8.118}$$

The only contact modes consistent with initial conditions are *separation* ($\ddot{\mathbf{p}}_{cn} > 0$) and left sliding. But it is easy to check that separation will not work. With no contact force, the net force in the normal direction is $-mg$, causing a *downward* acceleration.

That leaves us with left sliding. Coulomb's law gives

$$\mathbf{f}_t = \mu \mathbf{f}_n \tag{8.119}$$

or

$$\mathbf{f}_t + \mathbf{f}_n = f_n \begin{pmatrix} \mu \\ 1 \end{pmatrix} \tag{8.120}$$

The moment is

$$(\mathbf{p}_c - \mathbf{p}_0) \times (\mathbf{f}_n + \mathbf{f}_t) \tag{8.121}$$

$$= -l \begin{pmatrix} \cos\theta \\ \sin\theta \end{pmatrix} \times f_n \begin{pmatrix} \mu \\ 1 \end{pmatrix} \tag{8.122}$$

$$= l f_n (\mu \sin\theta - \cos\theta) \tag{8.123}$$

$$= l f_n \frac{1}{\cos\alpha} (-\cos(\alpha + \theta)) \tag{8.124}$$

where $\alpha = \tan^{-1} \mu$.

Substituting into equation 8.114:

$$\ddot{\mathbf{p}}_c = \frac{1}{m} f_n \begin{pmatrix} \mu \\ 1 \end{pmatrix} + \begin{pmatrix} 0 \\ -g \end{pmatrix} + \mathbf{0} + \frac{l^2 f_n}{I} \begin{pmatrix} -\sin\theta \\ \cos\theta \end{pmatrix} \frac{\cos(\alpha + \theta)}{\cos\alpha}$$

Now the left sliding mode specifies that $\ddot{p}_{cn} = 0$:

$$\frac{f_n}{m} - g + \frac{l^2 f_n}{I} \frac{\cos\theta}{\cos\alpha} \cos(\alpha + \theta) = 0 \tag{8.125}$$

Dynamics

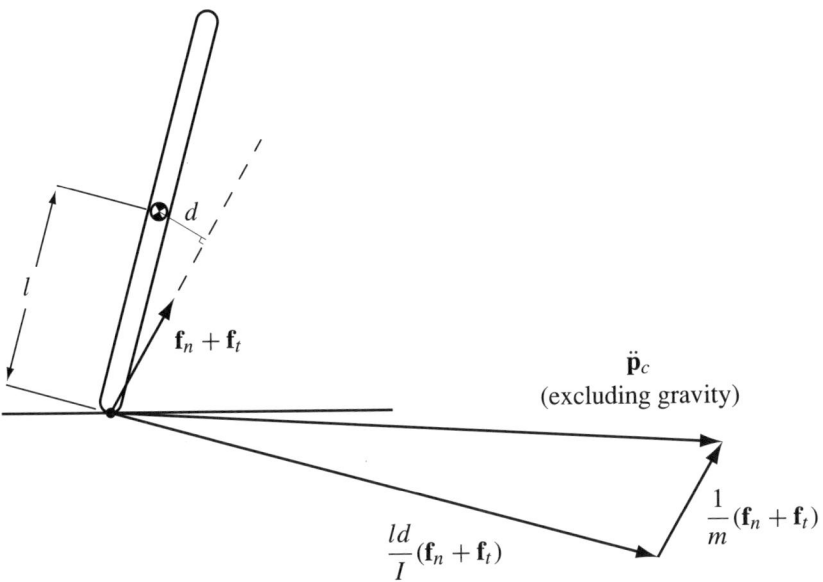

Figure 8.6
An example of frictional inconsistency. There is no choice of contact force that satisfies the assumptions of Newtonian mechanics, rigid bodies, and Coulomb friction.

or

$$f_n(\frac{1}{m} + \frac{l^2 \cos\theta}{I \cos\alpha} \cos(\alpha + \theta)) = g \qquad (8.126)$$

Now g is positive, and f_n is positive, so all values of $m, l, I, \alpha,$ and θ must satisfy

$$(\frac{1}{m} + \frac{l^2 \cos\theta}{I \cos\alpha} \cos(\alpha + \theta)) > 0 \qquad (8.127)$$

It turns out that for some θ this condition is violated by large values of μ, or small values of I. For example, if $m = I = 1, l = 4, \alpha = 30,$ and $\theta = 75$, then equation 8.127 is violated. There is no contact mode that works.

There is a simple geometrical explanation of the difficulty (figure 8.6). If we choose parameters placing the rod inside the friction cone, then the total force causes a positive moment at the center of mass. The acceleration of the contact point is the sum of a component due to the total force, which is away from the surface, and the total moment, which is into the surface. For small enough angular inertia, the component into the surface dominates. There is no force consistent with Coulomb's law that keeps the rod from penetrating the surface. However, this analysis also suggests a way out of the problem,

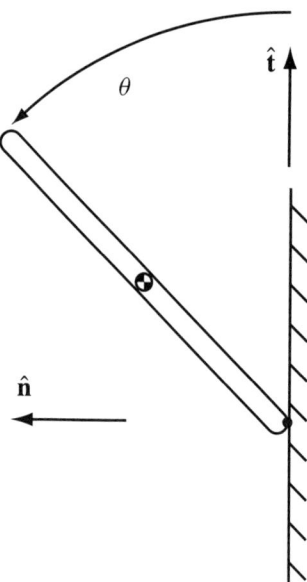

Figure 8.7
An example of frictional indeterminacy. If there is no contact force then the rod slips straight down without rotating. Or with a small contact force the rod will rotate about the contact point.

by interpreting the event as an impact. We shall see in chapter 9 that this approach works quite well.

Frictional indeterminacy

A small change in the problem yields multiple solutions. We change the direction of gravity, and start with a motionless rod (figure 8.7):

$$\mathbf{f}_a = \begin{pmatrix} -mg \\ 0 \end{pmatrix} \tag{8.128}$$

$$n_a = 0 \tag{8.129}$$

$$\dot{\mathbf{p}}_0 = \mathbf{0} \tag{8.130}$$

$$\dot{\theta} = 0 \tag{8.131}$$

All but three modes can easily be ruled out, leaving *separation*, *left-sliding*, and *fixed*. Separation does not work: we have $\mathbf{f}_t + \mathbf{f}_n = \mathbf{0}$, hence $\ddot{p}_{cn} = 0$, i.e. there is no normal acceleration.

For left sliding, there is a solution—if $\mathbf{f}_t = \mu \mathbf{f}_n = \mathbf{0}$, then we get $\ddot{p}_{ct} = -g$, i.e. the rod falls straight down, skimming the surface but with zero contact force.

For fixed contact, there is also a solution, depending on the physical parameters. We can simplify things by introducing a variable s:

$$s = \frac{f_t}{f_n} \tag{8.132}$$

For fixed contact, Coulomb requires that

$$-\mu \leq s \leq \mu \tag{8.133}$$

Now the total contact force can be written

$$\mathbf{f}_t + \mathbf{f}_n = f_n \begin{pmatrix} s \\ 1 \end{pmatrix} \tag{8.134}$$

Substituting into equation 8.114 with $\ddot{\mathbf{p}}_c = \mathbf{0}$,

$$\mathbf{0} = \frac{1}{m}\left(f_n \begin{pmatrix} s \\ 1 \end{pmatrix} + \begin{pmatrix} -mg \\ 0 \end{pmatrix}\right) + \frac{l}{I}\begin{pmatrix} -\sin\theta \\ \cos\theta \end{pmatrix} l \begin{pmatrix} \cos\theta \\ \sin\theta \end{pmatrix} \times f_n \begin{pmatrix} s \\ 1 \end{pmatrix}$$

Given the physical parameters m, I, g, l, and the initial angle θ, this yields two equations in F_n and s. It remains to be shown that values for these parameters exist yielding a positive value of F_n. The coefficient of friction can be arbitrarily large, and can always be chosen large enough for s to be in the range $[-\mu, \mu]$. The details are left for an exercise.

8.9 Graphical methods for the plane

Any of the methods described in chapter 5 to represent sets of forces in the plane—convex cones in wrench space, moment labeling, or force dual—provide a simple representation of the set of possible contact forces applied to a body. This gives an immediate solution to a typical statics problem: given an applied wrench $\mathbf{w}$ and a set of contact normals, to determine whether a body is going to move. The solution is to form the appropriate cone and ask whether the applied wrench $\mathbf{w}$ is balanced by a wrench in the cone. Let $\{\mathbf{c}_i\}$ be the set of contact wrenches. Then we can say that the object is in equilibrium under load $\mathbf{w}$ if and only if $\mathbf{w} \in \text{pos}(\{-\mathbf{c}_i\})$.

Perhaps more surprisingly, these same graphical methods will work for dynamics problems. Suppose a rigid planar body is initially at rest, but is being accelerated due to a total wrench $\mathbf{f} = (f_x, f_y, n_z)$, which we will refer to as the *dynamic load*. To simplify things, we will choose the origin to be the body center of mass, and also choose our unit of length to be the object radius of gyration. Let $\mathbf{a} = (a_x, a_y, \alpha_z)$ be the acceleration twist of the body. Then Newton's second law yields

$$\mathbf{a} = \begin{pmatrix} a_x \\ a_y \\ \alpha_z \end{pmatrix} = \frac{1}{m} \begin{pmatrix} f_x \\ f_y \\ n_z \end{pmatrix} \qquad (8.135)$$

With our choice of origin and unit of length, the dynamic load wrench is just the acceleration twist, scaled by the mass.

Now, suppose the dynamic load wrench $\mathbf{f}$ is actually composed of a some contact wrench $\mathbf{c}$ plus some applied wrench $\mathbf{w}$. Substituting above we obtain

$$\mathbf{a} = \frac{1}{m}(\mathbf{w} + \mathbf{c}) \qquad (8.136)$$

where $\mathbf{c}$ is the positive weighted sum of those contact normals or friction cone edges consistent with the contact mode

$$\mathbf{c} \in \mathrm{pos}(\{\mathbf{c}_i\}). \qquad (8.137)$$

Equation 8.136 defines a relation between a wrench $\mathbf{w}$ and a twist $\mathbf{a}$. The pair $\mathbf{w}, \mathbf{a}$ satisfies the relation if and only if there is some choice of contact forces satisfying Newton's second law. This relation applies to a point in wrench space and a point in twist space. Recall that the graphical methods work with *rays* in wrench and twist space, so we must extend the relation accordingly. We will say that the pair $\mathbf{w}, \mathbf{a}$ are related if and only if

$$\exists_{s \geq 0, s_i \geq 0} \; \mathbf{w} + \sum s_i \mathbf{c}_i = sm\mathbf{a} \qquad (8.138)$$

which is equivalent to

$$\mathbf{w} \in \mathrm{pos}(\{\mathbf{a}\} \cup \{-\mathbf{c}_i\}). \qquad (8.139)$$

Now we can develop the corresponding graphical methods. We will apply the force dual method. The moment labeling method is similar. Applying the dual map to equation 8.135 we obtain two equivalent expressions for the coordinates of the acceleration center:

$$\begin{pmatrix} -a_y/\alpha_z \\ a_x/\alpha_z \end{pmatrix} = \begin{pmatrix} -f_y/n_z \\ f_x/n_z \end{pmatrix} \qquad (8.140)$$

This equation, which follows directly from Newton's second law, shows that the acceleration center is obtained from the dynamic load by the force dual mapping. If we apply the force dual mapping to the applied wrench $\mathbf{w}$ and contact normals $\mathbf{c}_i$, then from equation 8.139 we obtain

Dynamics

$$\mathbf{w}' \in \text{conv}(\{\mathbf{a}'\} \cup \{-\mathbf{c}'_i\}) \tag{8.141}$$

An example problem is solved in the next section.

It should come as no surprise that we can also do the analysis by moment labeling. Given some candidate acceleration center, we apply the dual transformation to obtain the line of action of the dynamic load. We also plot the lines of action of the contact wrenches, but with the region labels negated. We intersect the labeled regions. The result describes the possible lines of action of an applied force consistent with the candidate acceleration center.

8.10 Planar multiple-contact problems

This section addresses the dynamics of a single rigid object in the plane, subject to several frictional contacts. In section 8.8 we described an algorithm for solving planar contact problems. We apply the method to planar single body multiple contact problems, and implement it with the force dual method. The first step is to enumerate the contact modes. Then, for each contact mode:

1. identify the set of acceleration centers;

2. identify the set of possible contact forces $\mathbf{c}_i$, negate them, and map them to their force duals;

3. form the convex hull of the acceleration centers and negated contact force duals, to obtain the set of applied wrenches represented as force duals.

When there are several contacts, the number of possible contact modes increases, so that the primary challenge is just keeping track of the different cases. We will adopt a convention for writing a contact mode. Suppose there are n point contacts on a rigid body in the plane. For the ith contact we will abbreviate the mode m_i as follows:

p   kinematically infeasible ("penetration")
s   separation
l   left sliding
r   right sliding
f   fixed

Then the contact mode of the body can be written as a string

$$m_1 m_2 \ldots m_n$$

This convention suggests that for n contacts there might be as many as 5^n different modes, but many of those are not kinematically consistent choices. For problems with only one

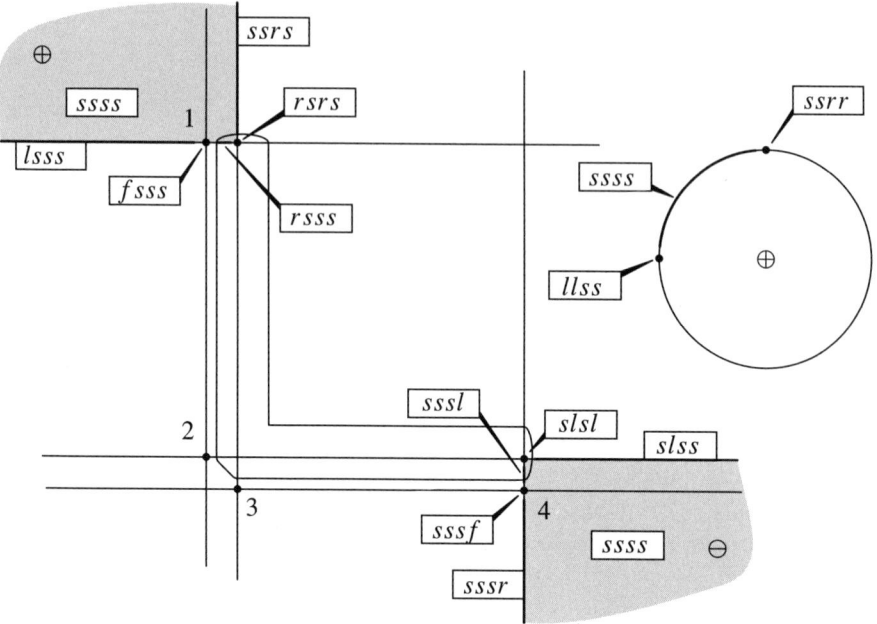

Figure 8.8
Reuleaux's method can be adapted to identify the kinematically feasible contact modes. f = fixed; s = separation; l = left sliding; r = right sliding.

moving body, we can use a variation on Reuleaux's method to enumerate the contact modes, and consider a number of possible modes that grows only as n^2.

The method is illustrated in figure 8.8, showing an allen wrench initially at rest in the corner of a tray. We construct the contact normals, and the contact tangents, at each contact. These lines divide the space of signed rotation centers up into polygonal regions, line segments, and points. If we use the term *cell* to refer to a region, segment, or point, then each cell corresponds to a different contact mode. For each cell, we simply write the mode. Excluding the penetrating modes yields the set of all kinematically feasible contact modes.

We can now proceed with the dynamic analysis, with either moment labeling or force dual representations, as in section 8.9. Each contact mode is already represented by a convex set of acceleration centers. To this set we add the force duals of all possible contact wrenches negated $\{-\mathbf{c}'_i\}$ and take the convex hull.

Figure 8.9 shows the constructions for the contact mode "rsrs", using the acceleration centers obtained in figure 8.8. Using the wrench mass center as the origin, and the radius of

Dynamics

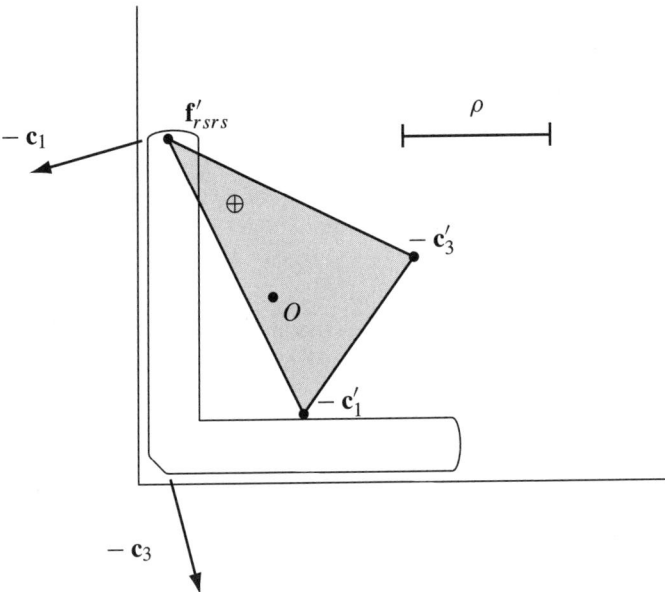

Figure 8.9
Analysis of mode "rsrs" (counterclockwise rotation with contact at c_1 and c_3.)

gyration ρ as the unit of length, this acceleration center is the dual $\mathbf{f}'$ of the dynamic load wrench. Since we have right sliding at contacts 1 and 3, the contact wrenches $\mathbf{c}_1$ and $\mathbf{c}_3$ lie on the left edges of their respective friction cones. Since we have separation at contacts 2 and 4, they contribute no forces. So the duals of the negated contact forces are plotted $-\mathbf{c}'_1$ and $-\mathbf{c}'_3$. Finally, the convex hull is taken, yielding a triangle in force dual space. Any applied force whose dual falls in this triangle can cause the wrench to move in mode "rsrs".

8.11 Bibliographic notes

The development of rigid body dynamics was adapted from Symon's text (1971). The treatment of the dynamics of rigid bodies with frictional contact was adapted from (Erdmann, 1994). Painlevé (1895) was the first to notice the difficulties arising from Coulomb friction. For further work on the problem see for example (Baraff, 1990, 1993), (Trinkle et al., 1997), and (Stewart, 1998).

Exercises

Exercise 8.1: The purpose of this exercise is to address false intuitions about the motion of a tumbling body. Also, it is fun. Take a hardcover book (not this one!), and fasten it closed with a rubber band. Imagine a coordinate frame at the center of mass, with $\hat{\mathbf{x}}$ orthogonal to the covers of the book, $\hat{\mathbf{z}}$ parallel to the spine of the book, and $\hat{\mathbf{y}}$ constructed to give a right-handed coordinate frame. Then $\hat{\mathbf{x}}$ is the major principal axis, $\hat{\mathbf{z}}$ the minor principal axis, and $\hat{\mathbf{y}}$ the intermediate principal axis. With some practice, you should be able to toss the book in the air, and achieve a nearly-constant angular velocity about either the $\hat{\mathbf{x}}$-axis or the $\hat{\mathbf{z}}$-axis. Now try to do the same for the $\hat{\mathbf{y}}$-axis. It is virtually impossible. What would the motion look like, if the angular velocity were constant? If the binding of the book is initially on your left, for example, would it be possible for it to end on your right?

Exercise 8.2: Consider the book-throwing experiment of exercise 8.1. Show that if you could give the book an initial angular velocity that is *exactly* aligned with the $\hat{\mathbf{y}}$-axis, the angular velocity vector would be constant. Show that a small deviation will result in an angular velocity vector that alternates between nearly parallel to $\hat{\mathbf{y}}$ and nearly parallel to $-\hat{\mathbf{y}}$.

Exercise 8.3: Prove theorems 8.1 and 8.2.

Exercise 8.4: Prove theorem 8.3.

Exercise 8.5: Find the inertia matrix for a right circular cylinder of radius r, height h, and total mass M, of uniform density.

Exercise 8.6: Construct the inertia matrix, and sketch the inertia ellipsoid, for a pencil, and for a coin. You may model them as right circular cylinders, using dimensions measured from any convenient pencil and coin, and use the result of exercise 8.5. You need not bother weighing them, since the scale of the inertia ellipsoid is arbitrary.

Exercise 8.7: Euler's equations can be awkward when dealing with degenerate objects. Suppose two point masses are connected by a weightless rod. Then rotation of the rod about its own axis is indeterminate, since it does not involve the motion of any mass. Let the distance between the two masses be two, and suppose the rod is initially aligned with the z-axis. We give the rod an initial angular velocity $\boldsymbol{\omega} = (0, 1, 1)$, then let it rotate freely. Apply Euler's equations and explain the results.

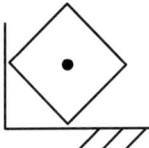

Figure 8.10
Drawing for exercise 8.13.

Exercise 8.8: Repeat the dynamic analysis of section 8.8, using the moment labeling method.

Exercise 8.9: Repeat exercise 8.8, using the force dual method.

Exercise 8.10: Complete the example of frictional ambiguity, by producing meaningful physical parameters, solving for F_n and s, and showing that all the necessary conditions are satisfied.

Exercise 8.11: Repeat the analysis of the allen wrench in contact mode "rsrs" using the moment labeling method.

Exercise 8.12: Do the analysis of the allen wrench in contact mode "rsss", using either the force dual or the moment labeling method.

Exercise 8.13: Use the force dual method to analyze the block in the corner shown in figure 8.10. The block is one by one, the radius of gyration is one, and the coefficient of friction is 0.25. Use Reuleaux's method to identify the contact modes. For each contact mode, form the possible dynamic load duals, the negated contact force duals, and use convex hull to find the possible applied force duals.

9 Impact

This chapter develops basic ideas for modeling frictional impact, and illustrates by way of two examples. Modeling impact is so fraught with complexities and subtleties that we must stick to the simplest models involving rigid bodies in the plane.

What happens when two rigid bodies collide? There is a discontinuity in the body velocities, implying infinite accelerations and infinite forces. Some matters are simplified. For example, the sliding rod inconsistency of chapter 6 can be resolved using impact. On the other hand, widely accepted models of impact exist only for the simplest cases. And even the simplest cases of frictional impact can cause difficulty.

To avoid dealing directly with infinite forces, we use *impulse* instead. Recall from chapter 8:

$$\mathbf{F} = \frac{d\mathbf{P}}{d\mathbf{t}} \tag{9.1}$$

$$\mathbf{N} = \frac{d\mathbf{L}}{d\mathbf{t}} \tag{9.2}$$

Integrating:

$$\Delta\mathbf{P} = \int_{t_0}^{t_1} \mathbf{F}\, dt \tag{9.3}$$

$$\Delta\mathbf{L} = \int_{t_0}^{t_1} \mathbf{N}\, dt \tag{9.4}$$

where $\Delta\mathbf{P}$ is the change in momentum $\mathbf{P}_1 - \mathbf{P}_0$, $\Delta\mathbf{L}$ is the change in angular momentum $\mathbf{L}_1 - \mathbf{L}_0$, $\int \mathbf{F}\, dt$ is the *impulse*, and $\int \mathbf{N}\, dt$ is the *impulsive moment*. These are the so-called *impulse-momentum equations*, which are nothing more than Newton's laws, presented in a form suitable for impact problems. Now all we need is some constitutive law that determines the total impulse, and impulsive moment, as a function of the bodies' physical parameters and initial state. Unfortunately there is no simple law that does the trick. Our goal will be narrow: to address the simplest model sufficient for planar rigid-body impact problems, and explore the implications.

9.1 A particle

Suppose a particle in the plane impacts on a horizontal surface, without friction. We will model the impact by considering an impulse $\mathbf{I}$ acting over some interval of time $[t_0, t_1]$, and take the limit as t_1 approaches t_0. Thus the particle velocity changes discontinuously from $\mathbf{v}_0$ to $\mathbf{v}_1$:

$$\Delta \mathbf{v} = \frac{1}{m}\Delta \mathbf{P} = \frac{1}{m}\mathbf{I} \tag{9.5}$$

where

$$\Delta \mathbf{v} = \mathbf{v}_1 - \mathbf{v}_0 \tag{9.6}$$

For the frictionless case the tangential impulse I_t is zero, so

$$v_{1t} = v_{0t} \tag{9.7}$$

$$v_{1n} = v_{0n} + \frac{1}{m}I_n \tag{9.8}$$

So all we require is some constitutive law giving the normal impulse I_n. There are two special cases: *plastic impact*

$$I_n = -mv_{0n} \tag{9.9}$$

$$\rightarrow v_{1n} = 0 \tag{9.10}$$

and *elastic impact*

$$I_n = -2mv_{0n} \tag{9.11}$$

$$\rightarrow v_{1n} = -v_{0n} \tag{9.12}$$

Newton hypothesized a continuum from plastic to elastic, defined by a *coefficient of restitution*

$$e = -\frac{v_{1n}}{v_{0n}} \tag{9.13}$$

so that plastic impact corresponds to $e = 0$, and elastic impact corresponds to $e = 1$.

A competing definition of the coefficient of restitution was given by Poisson. Note that, since the contact force is non-negative, the particle's normal velocity is monotone non-decreasing. So we can divide the collision into two stages.

$v_n < 0$ compression

$v_n = 0$

$v_n > 0$ restitution

Impact

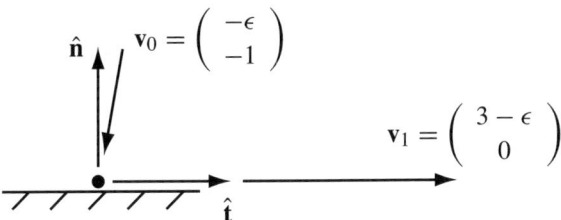

Figure 9.1
An example showing the need to identify different stages of impact as the contact mode changes.

We divide the impulse into two parts, I_c applied during compression, and I_r applied during restitution. Poisson's hypothesis is that the ratio of these two parts is governed by material properties:

$$e = \frac{I_r}{I_c} \tag{9.14}$$

Like Newton's approach, this yields $e = 1$ for elastic impact, and $e = 0$ for plastic impact. And for the frictionless point, the two approaches yield identical results. But for some problems Newton's definition is unworkable, so we will adopt Poisson's hypothesis.

Returning to the problem at hand, the approach is quite simple. Given the particle mass m, initial velocity $\mathbf{v}_0$, and a coefficient of restitution e, we first obtain the impulse during compression:

$$I_c = mv_{0n} \tag{9.15}$$

Then we multiply by the coefficient of restitution to obtain the impulse during restitution:

$$I_r = emv_{0n} \tag{9.16}$$

obtaining a total normal impulse:

$$I = I_c + I_r = (1+e)mv_{0n} \tag{9.17}$$

and a final normal velocity:

$$v_{1n} = -ev_{0n} \tag{9.18}$$

The tangential velocity is unchanged.

FRICTION: A BAD MODEL

Now consider the same problem, but with friction. First, we need a law to determine the tangential impulse I_t. As a first try, we might choose a straightforward extension of Coulomb's law:

$$I_t = s I_n \tag{9.19}$$

where

$$\begin{aligned} s &= \mu & \text{if } v_{0t} &< 0 \quad \text{(left-sliding)} \\ -\mu &\leq s \leq \mu & \text{if } v_{0t} &= 0 \quad \text{(fixed)} \\ s &= -\mu & \text{if } v_{0t} &> 0 \quad \text{(right-sliding)} \end{aligned}$$

but this doesn't work very well. Figure 9.1 shows an example. Suppose that we have a plastic collision ($e = 0$) with a very high coefficient of friction ($\mu = 3$). If the particle strikes in the normal direction ($v_{0t} = 0$), it comes to rest, as expected. Suppose however, that the particle starts with unit normal velocity and a very small leftward tangential velocity:

$$v_{0n} = -1 \tag{9.20}$$
$$v_{0t} = -\epsilon \tag{9.21}$$

Because it is a plastic impact, the normal impulse would be just enough to bring the normal velocity to zero:

$$I_n = m \tag{9.22}$$

Since the particle struck in left-sliding mode, the tangential impulse would be

$$I_t = \mu I_n = 3m \tag{9.23}$$

The final velocity, then, would be

$$v_{1n} = 0 \tag{9.24}$$
$$v_{1t} = 3 - \epsilon \tag{9.25}$$

yielding a gain in kinetic energy.

A BETTER MODEL

It is not hard to see what went wrong with our first attempt at a frictional impact law. We applied the left-sliding rule over the entire duration of the impact, even though the particle switched to right-sliding during the impact. Just as we divided the impact into the stages

Impact

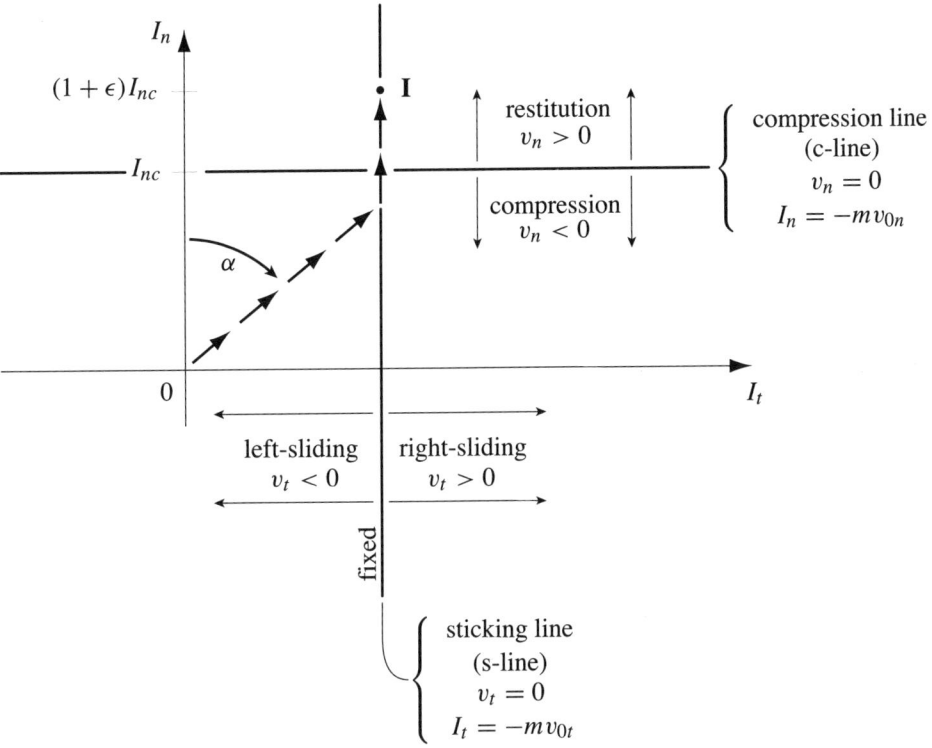

Figure 9.2
Routh's use of impulse space to analyze planar impact of a particle.

of *compression* and *restitution*, we will also divide the impact according to the contact mode, into the stages of *left sliding*, *fixed*, and *right sliding*. Note that some collisions might involve only one or two contact modes, though, so we do not necessarily see all three stages in every collision. In the example of figure 9.1, for example, the particle would start in the *left sliding* stage, then switch to the *fixed* stage, and come to rest.

In order to keep the stages right, Routh described a simple graphical method, which plots the impulse as it accumulates during the collision. Consider the *impulse space* of figure 9.2. The stages are easily described in impulse space. First consider Poisson's distinction between compression and restitution. The switch from compression to restitution occurs when the normal velocity v_n is zero. This gives us a line in impulse space: $I_n = -mv_{0n}$. Below this line we have compression; above it restitution. We call it the compression line, or *c-line*. Similarly, the sliding modes are determined by the condition of zero tangential velocity. This gives a second line in impulse space: $I_t = -mv_{0t}$. Inside

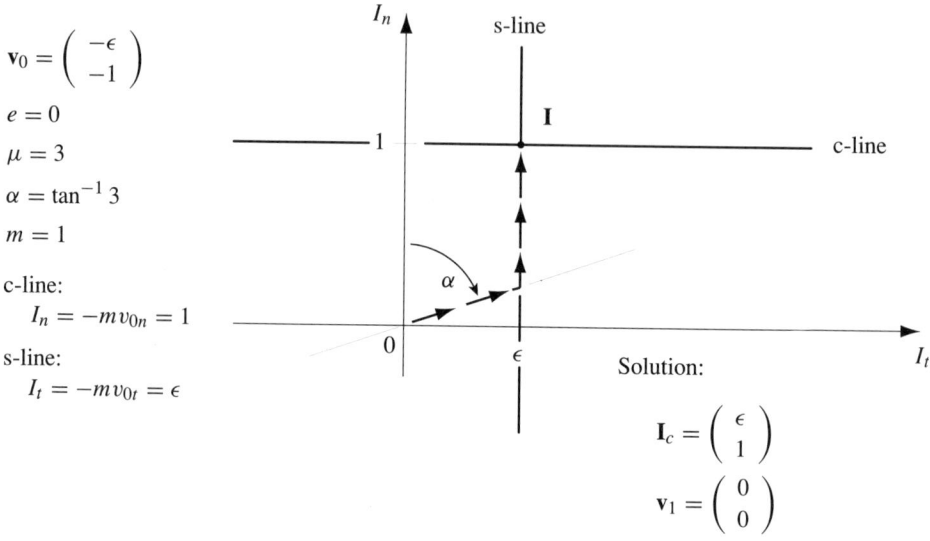

Figure 9.3
Routh's method applied to the problem of figure 9.1.

the line we have forward sliding, on the line we have fixed, and outside the line we have reverse sliding. This line is called the sticking line, or *s-line*.

Using impact space, we easily obtain a more sensible analysis (figure 9.3) of a particle hitting a surface with high friction. As before, we will start with the particle drifting slightly to the left. Initially there is no impulse, so the impulse starts at **0**, and builds up, along the right edge of the friction cone. However, it soon intersects the s-line, where the tangential velocity is zero. Now the impulse will grow along the sticking line. No other choice gives an incremental impulse consistent with the contact mode. The impulse continues to grow until it strikes the c-line, indicating the normal velocity is about to change sign, changing from compression to restitution. We note the normal impulse accumulated so far, I_{nc}, and calculate the total normal impulse $(1 + e)I_{nc}$. Now the total impulse grows, along the sticking line, until we hit a line $I_n = (1+e)I_{nc}$, indicating that the impact is over. However, in this case the coefficient of restitution is $e = 0$, so the impact ends at the compression line. So the particle started in left-sliding and compression, then switched to fixed and compression. The particle ended at rest. If the collision had been an elastic collision, the particle would have rebounded straight up.

Impact

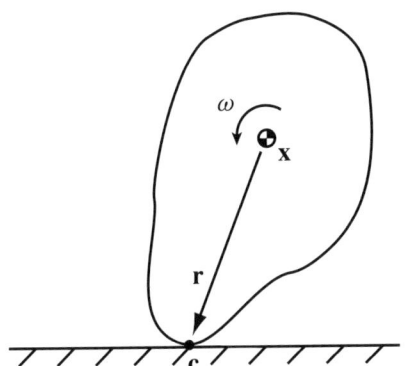

Figure 9.4
Notation for rigid body impact.

9.2 Rigid body impact

We can apply Routh's impulse space to analyze a rigid body impact. Assume a rigid body is striking an immobile rigid half plane (figure 9.4). To simplify the analysis, we express the impulse-momentum equations in terms of the contact point. First, we observe some kinematic relations

$$\mathbf{c} = \mathbf{x} + \mathbf{r} \tag{9.26}$$
$$\dot{\mathbf{c}} = \dot{\mathbf{x}} + \dot{\mathbf{r}} \tag{9.27}$$
$$= \dot{\mathbf{x}} + \omega \times \mathbf{r} \tag{9.28}$$
$$\Delta\dot{\mathbf{c}} = \Delta\dot{\mathbf{x}} + \Delta\omega \times \mathbf{r} \tag{9.29}$$

where $\mathbf{c}$ is the contact point, $\mathbf{x}$ is the center of mass, $\mathbf{r}$ is the radius vector from the center of mass to the contact point, and ω is the angular velocity.

Now we can write the impulse-momentum laws:

$$m\Delta\dot{\mathbf{x}} = \mathbf{I} \tag{9.30}$$
$$\rho^2 m \Delta\omega = \mathbf{r} \times \mathbf{I} \tag{9.31}$$

Substituting into the expression for $\Delta \dot{\mathbf{c}}$ yields:

$$\Delta \dot{\mathbf{c}} = \frac{1}{m}\mathbf{I} + \frac{1}{\rho^2 m}(\mathbf{r} \times \mathbf{I}) \times \mathbf{r} \tag{9.32}$$

$$= \frac{1}{m}\mathbf{I} - \frac{1}{\rho^2 m}\mathbf{r} \times (\mathbf{r} \times \mathbf{I}) \tag{9.33}$$

$$= \frac{1}{\rho^2 m}\left(\rho^2 I_3 - R^2\right)\mathbf{I} \tag{9.34}$$

where I_3 is the three by three identity matrix, and R is the cross-product matrix

$$R = \begin{pmatrix} 0 & 0 & r_n \\ 0 & 0 & -r_t \\ -r_n & r_t & 0 \end{pmatrix} \tag{9.35}$$

Substituting above and expanding, we obtain

$$\begin{pmatrix} \Delta \dot{c}_t \\ \Delta \dot{c}_n \end{pmatrix} = \frac{1}{\rho^2 m} \begin{pmatrix} (\rho^2 + r_n^2) & -r_t r_n \\ -r_t r_n & (\rho^2 + r_t^2) \end{pmatrix} \begin{pmatrix} I_t \\ I_n \end{pmatrix} \tag{9.36}$$

As with the particle, the impulse-momentum equations are linear, so the sticking line and compression line are described by lines in Routh's impulse space. The sticking line is defined by the equation $\dot{c}_t = 0$,

$$\dot{c}_{0t} + \frac{\rho^2 + r_n^2}{\rho^2 m} I_t - \frac{r_t r_n}{\rho^2 m} I_n = 0 \tag{9.37}$$

Similarly, the compression line is defined the equation $\dot{c}_n = 0$,

$$\dot{c}_{0n} - \frac{r_t r_n}{\rho^2 m} I_t + \frac{\rho^2 + r_t^2}{\rho^2 m} I_n = 0 \tag{9.38}$$

As with the particle, we can use the sticking line and compression line to divide the impact into stages, applying Coulomb's law and Poisson's restitution to analyze the impact.

EXAMPLE 1

Figure 9.5 shows a rigid rectangle striking an immobile rigid half-plane. The line of sticking is

$$-1 + 5I_t + 2I_n = 0 \tag{9.39}$$

and the line of compression is

$$-1 + 2I_t + 2I_n = 0 \tag{9.40}$$

Impact

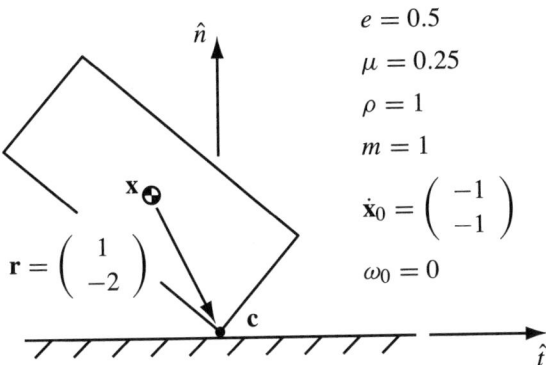

Figure 9.5
Example 1

The progress of the impact is traced in the impulse space of figure 9.6:

1. Left sliding, compression. Impulse accumulates along the right edge of the friction cone.
2. After hitting the sticking line, the accumulated impulse will cross into right sliding, so further accumulations of impulse are along the left edge of the friction cone.
3. Right sliding, compression.
4. After hitting the compression line, we switch to the restitution stage. Note the normal impulse I_{nc}.
5. Right sliding, restitution.
6. When $I_n = 1.5 I_{nc}$, collision is over.

EXAMPLE 2

Our second example is the sliding rod problem treated in section 8.8, redrawn in figure 9.7. The sticking line is

$$-1 + 17 I_t - 4 I_n = 0 \tag{9.41}$$

and the compression line is

$$-4 I_t + 2 I_n = 0 \tag{9.42}$$

Notice that the compression line passes through the origin, because there is no initial normal velocity. As a result, it is hardly clear that this should be regarded as a collision, but we will continue, and return to the question later.

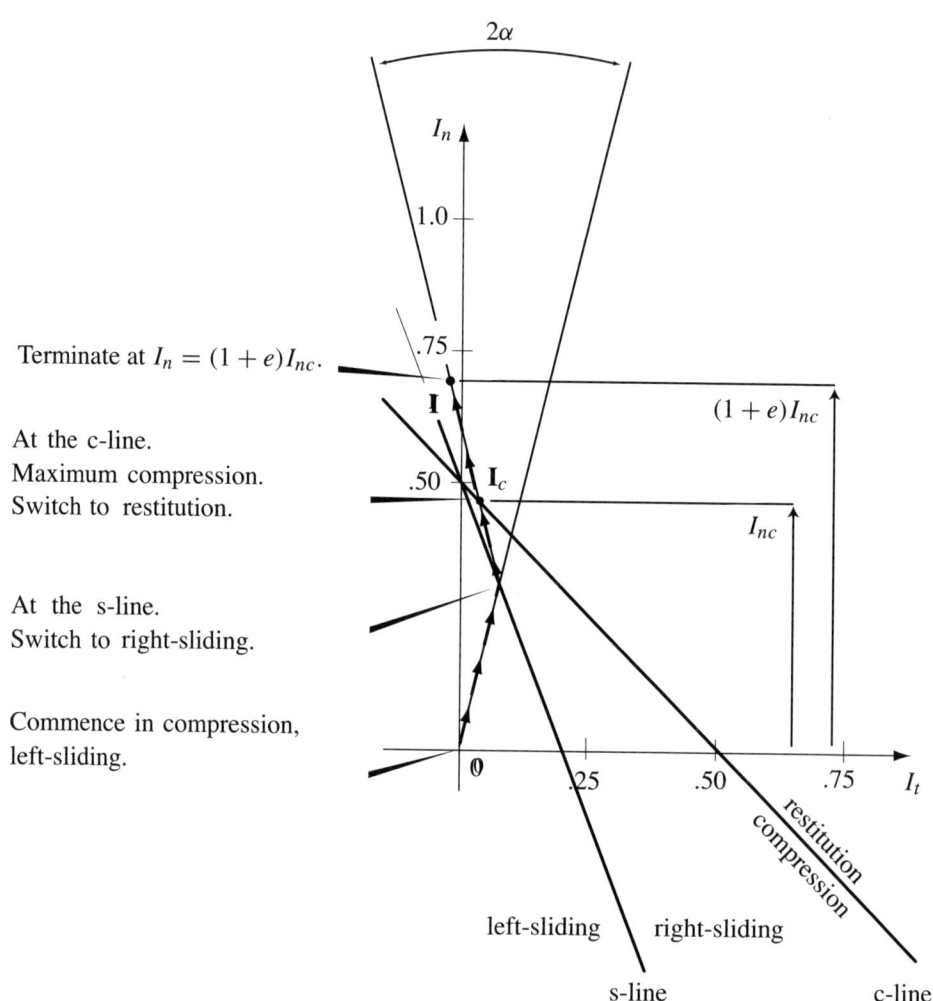

Figure 9.6
Routh's method applied to Example 1.

Figure 9.8 shows the impulse space for the sliding rod problem. Initially the rod is in left sliding compression. When it strikes the sticking line, there are three possibilities to consider. It cannot continue along the right edge of the friction cone, because it would cross the sticking line into right sliding mode. It also cannot switch to the left edge of the friction cone, which would put it back in left sliding mode. The only consistent choice is to follow the sticking line, which falls in the interior of the friction cone.

Impact

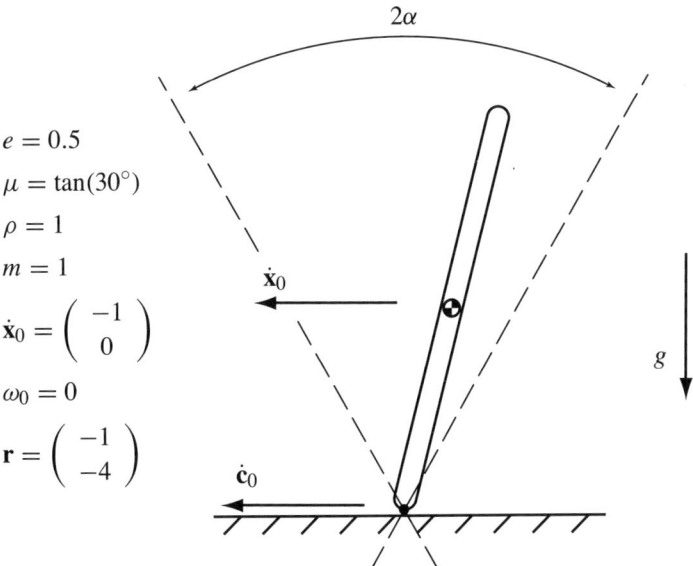

Figure 9.7
A sliding rod problem with no finite force solution.

Then it crosses the compression line, switching to restitution mode, and continues until the impact is finished.

Routh's approach offers a solution, which is consistent with the impulse-momentum equations, with Coulomb's law of sliding friction, and with Poisson's restitution hypothesis. The only question is whether it makes sense to use an impact solution when there is no initial normal velocity, a *tangential impact*. There are two arguments in favor. First, in this case at least, there are *no* alternative solutions—there is no set of finite contact forces that are consistent with the relevant axioms. Second, the phenomenon is not wholly unfamiliar to those who wear rubber-soled shoes on clean floors. Occasionally, in such circumstances, one seems to trip over an invisible obstacle.

As a final remark, note that Newton's definition of restitution does not behave very well for tangential impact. Since the initial normal velocity is zero, Newtonian restitution would treat elastic impact and plastic impact identically. Poisson's definition fits tangential impact very neatly.

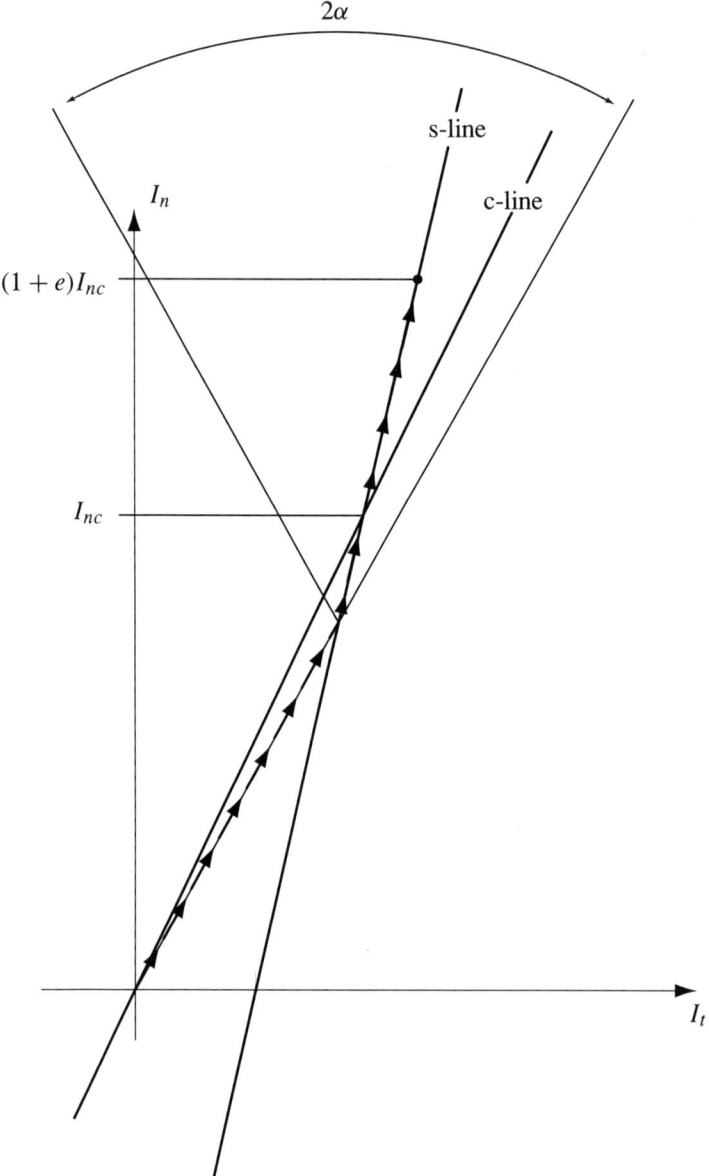

Figure 9.8
Routh's method gives an impulsive force solution to the sliding rod problem.

9.3 Bibliographic notes

The analysis of impact is adapted from Wang's work (1992), which was based on Routh's method (1913). Stewart (1998) unifies the treatment of impact with finite forces, and shows that the tangential collision solutions occur precisely when the finite force solutions fail.

There is disagreement about how rigid body impact should be modeled. Many different laws have been proposed, but none are entirely satisfactory. Chatterjee and Ruina (1998) use impulse space to analyze and compare a number of different impact laws, and introduce a new law to obtain the best features while avoiding the defects. Also see (Stewart, 2000) for a survey of modern work on rigid body frictional impact.

Exercises

Exercise 9.1: Repeat the analysis of example 1, but with an initial angular velocity of 1 radian per second. Give the equations for the s-line and the c-line, and construct the impulse space diagram to determine the total impulse.

Exercise 9.2: Repeat the analysis of example 1, but with a radius of gyration of 2. Give the equations for the s-line and the c-line, and construct the impulse space diagram to determine the total impulse.

Exercise 9.3: In example 1 the mode of impact is called "reverse sliding" because the sliding switches direction, in this case from left sliding to right sliding. For a large enough coefficient of friction μ, the sliding will be stopped but not reversed. Find the smallest μ that will stop but not reverse the sliding for example 1, and construct the impulse space diagram.

10 Dynamic Manipulation

The taxonomy of manipulation presented in chapter 1 starts with kinematic manipulation, proceeds through static and quasistatic manipulation, and ends with dynamic manipulation. To some extent, robotics research has followed the same progression. There are numerous studies, both theoretical and empirical, that address the problems of kinematic manipulation. Quasistatic manipulation has received less attention, but is still fairly mature. Dynamic manipulation is in a more formative stage. Some far-thinking researchers started on dynamic manipulation many years ago, laying the groundwork and developing many of the key ideas. But the mechanics of dynamic manipulation is still being developed.

This chapter surveys a few examples of dynamic manipulation drawn from the research literature. The survey is organized as a progression from the barely dynamic to the gloriously dynamic. At the "barely dynamic" end of the spectrum is *quasidynamic manipulaton*, where the dynamic periods are so brief that we can model them as vanishingly small. Next, we will consider examples of *briefly dynamic* manipulation, which are charactized by brief excursions into dynamics, surrounded by more conservative kinematic or quasistatic periods. Finally we consider *continuously dynamic* manipulation, where dynamic processes persist for an indefinitely long period of time.

10.1 Quasidynamic manipulation

Quasidynamic analysis is intermediate between quasistatic and dynamic. Suppose that a manipulation task involves occasional brief periods when there is no quasistatic balance. The task is then governed by Newton's laws. But in some instances, these periods are so brief that the accelerations do not integrate to significant velocities. Momentum and kinetic energy are negligible. Restitution in impact is negligible. It is as if a viscous ether is constantly damping all velocities and sucking the kinetic energy out of all moving bodies. We can analyze such a system by assuming it is at rest, calculating the total forces and body accelerations, then moving each body some short distance in the corresponding direction.

Quasidynamic analysis is simpler than dynamic analysis, and is accurate enough in some cases. The clearest advantage is that the state of a mechanical system is its configuration—half the state variables required for a dynamic analysis.

The rest of this section illustrates quasidynamic analysis by the example of tray tilting.

Tray tilting

This section presents an example of quasidynamic manipulaton, and shows how the dynamic analysis of section 8.10 can be used to manipulate the allen wrench by tilting the

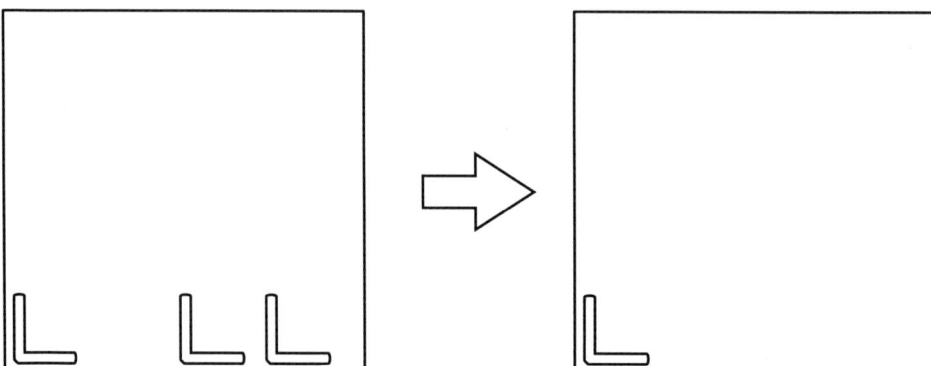

Figure 10.1
The goal is to move the wrench to the lower left corner from any position along the bottom wall.

tray. Besides the quasidynamic assumption, we will also assume friction with the floor of the tray is negligible. As our task, we will assume that the wrench is initially positioned somewhere along the bottom wall of the tray, with its short end aligned with the wall (figure 10.1). The goal is to translate it into the lower left corner.

The first question is: what motions of the allen wrench can be achieved by tilting the tray? First we will look at the goal configuration in the lower left corner. Section 8.10 analyzes the dynamics of this configuration. Figure 8.8 identifies all the feasible contact modes, and figure 8.9 analyzes the contact mode "rsrs", in which the wrench rotated counterclockwise in two point contact. Now consider the question, can mode "rsrs" be produced by tilting the tray? When we tilt the tray, we obtain a gravitational force through the center of mass. The corresponding force dual is at infinity. Unfortunately the triangle of force duals shown in figure 8.9, corresponding to mode "rsrs", does not intersect the line at infinity. Hence it is impossible to produce "rsrs" by tilting the tray.

By repeating this analysis for each contact mode in figure 8.8 we can identify every motion that can be produced by tilting the tray. Figure 10.2 analyzes mode "ssrr"—sliding to the right in two point contact. In section 8.10 we used the force dual method. We now switch to moment labeling, since so much of the force dual action would be at infinity. For each contact mode:

1. identify the set of acceleration centers, and apply the dual transform to obtain a moment labeling representation of the dynamic loads $\mathbf{f}$;

2. for each possible contact force $\mathbf{c}_i$, negate it and construct the moment labeling representation;

Dynamic Manipulation

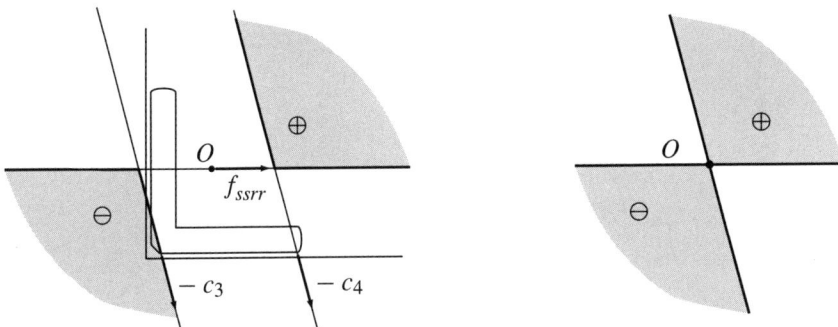

Figure 10.2
Analysis of mode "ssrr" (sliding to the right in two point contact) by moment labeling.

3. intersect all the positive-labeled regions and all the negative-labeled regions, to obtain the set of applied wrenches represented by moment labeling.

For the case at hand, the single acceleration center is at infinity (figure 8.8). Its dual, the dynamic load, is a rightward force acting through the center of mass (figure 10.2). For right sliding, the two possible contact forces are the left edges of the friction cones at the two contact points. We negate them, label regions, and intersect the labeled regions. The result is the set of all wrenches that would lead to mode "ssrr".

To identify wrenches that can be produced by tilting the tray, we add a single point labeled ± at the center of mass, and take the convex hull. The result is a simple cone at the center of mass, indicating which tray tilt directions would cause mode "ssrr"—rightward sliding in two point contact.

Repetition of this analysis for every contact mode yields only four modes that are feasible by tilting the tray: right sliding along the bottom wall ("ssrr"), left sliding up the left wall "llss", separation ("ssss"), and immobility ("ffff"). Constructing the cone for each mode gives a complete map of the feasible motions from this configuration (figure 10.3).

Figure 10.4 shows a similar map, for initial configuration of the wrench against the lower wall. The details are left as exercises. As with the configuration in the corner, there are only four feasible contact modes, one of which moves the wrench toward the corner.

Figure 10.4 also shows a combined map, which identifies all possible actions given that the wrench is either against the lower wall or in the corner. From this map we can find a range of tilt angles that accomplish the goal, no matter which initial condition applies.

228   Chapter 10

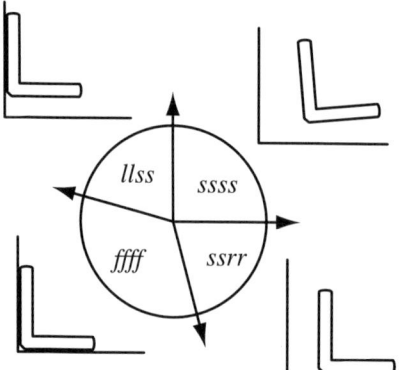

Figure 10.3
A map showing every motion of the wrench that can be produced by tilting the tray, assuming it starts in the lower left corner.

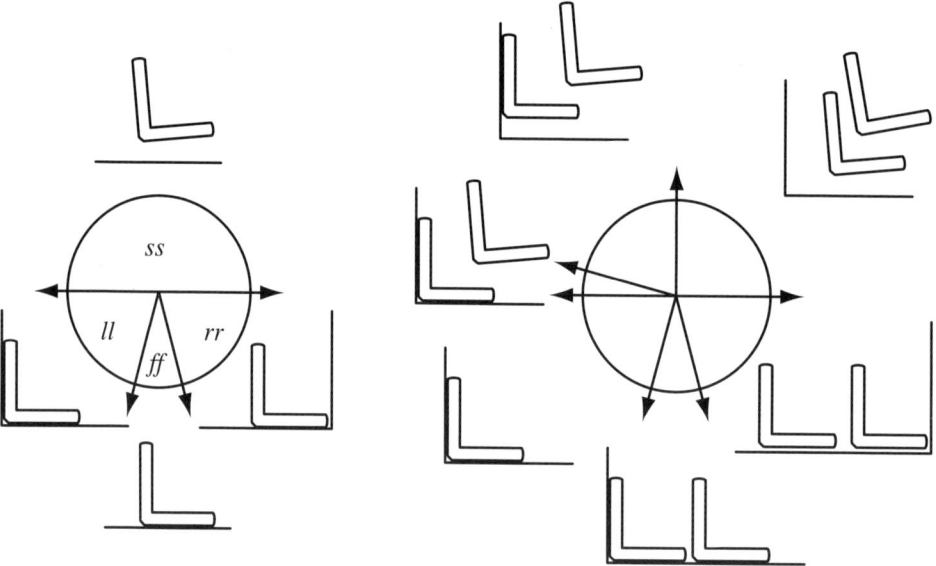

Figure 10.4
A map showing every motion of the wrench that can be produced by tilting the tray, assuming it starts against the lower wall. And a map showing every motion assuming a start against the lower wall *or* in the lower left corner.

10.2 Briefly dynamic manipulation

Some dynamic tasks include kinematic or quasistatic periods, so that every dynamic phase has a finite time horizon. An obvious example would be a high speed grasping motion which might depend on impact and dynamic forces for its effect, but which ends with the object in a form closure grasp. A much less obvious example is juggling three balls with two hands. The deficit of hands dictates that at all times at least one ball is in flight. However, if we adopt the perspective of a single ball, we see that the ballistic phase is interleaved with an in-hand phase which might be kinematic or quasistatic. Such a juggle can be modeled as three briefly dynamic processes running in parallel, sharing the hands. To consider such a task in more detail, the rest of this section focuses on Shannon's juggler.

Shannon's juggler

Claude Shannon is generally credited with building the first juggling machine. It juggled three balls using two hands. The machine used a technique known as *bounce juggling*, which usually means that each ball bounces on the floor between throw and catch. The design (figure 10.5) employs two cups, padded with an energy absorbing material, mounted at either end of a roughly horizontal rocker arm. The arm oscillates about its center. Each cup is mounted so that at the zenith of its travel, the ball is tossed out of the cup, falls to a drumhead below, and bounces into the opposite cup, as that cup nears its nadir. The throwing motion is simple, because the hand does not have to produce precisely the desired motion of the ball, nor is there an elaborate mechanism to release the ball at precisely the right time.

Why does it work? The motion of the machine is unvarying, so it would seem there is no mechanism to detect and correct deviations of the balls from the desired paths. The energy absorbing lining of the cups is the key. The ball lands in the cup and rolls to the bottom, so that the variations from one cycle are eliminated before the next cycle begins. Eliminating uncertainty is often linked with eliminating kinetic energy. The link is given by Liouville's theorem. We can represent the state of an n degree of freedom dynamic system by its *phase*: n configuration variables and n momentum variables. An uncertain system can be represented by an uncertainty cloud in phase space, the locus of all states the system could be in. Liouville's theorem states that for a Hamiltonian system (which would include a passive energy-conserving mechanical system) the volume of that uncertainty cloud in phase space remains constant. The uncertainty cloud will generally move around in phase space and change shape, but it cannot shrink. So in order to stabilize the flight of the ball, we must violate the assumptions of Liouville's theorem by avoiding passive energy-conserving systems. One way to do it is to cover the cups with energy-absorbing foam.

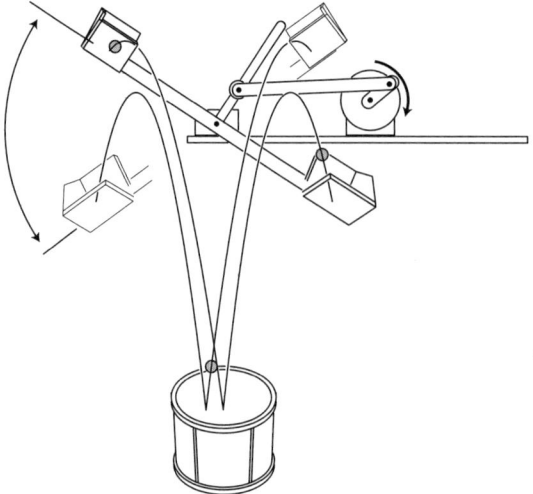

Figure 10.5
Schaal's version of the Shannon juggler. Adapted from Schaal et al. (1992).

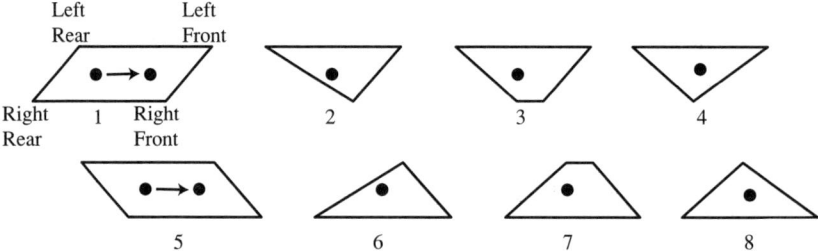

Figure 10.6
For a statically stable gait the center of mass is always above the support polygon. The feet of this quadruped are at the vertices of the polygon, and the center of mass is above the dot. Adapted from Raibert (1986) who adapted it from McGhee and Frank (1968).

10.3 Continuously dynamic manipulation

Some tasks are continuously dynamic, and have to work without resorting to rest periods of kinematic or quasistatic manipulation. Juggling *without* energy absorbing foam would be one example, but we will instead consider dynamic locomotion: running and hopping.

Dynamic Manipulation

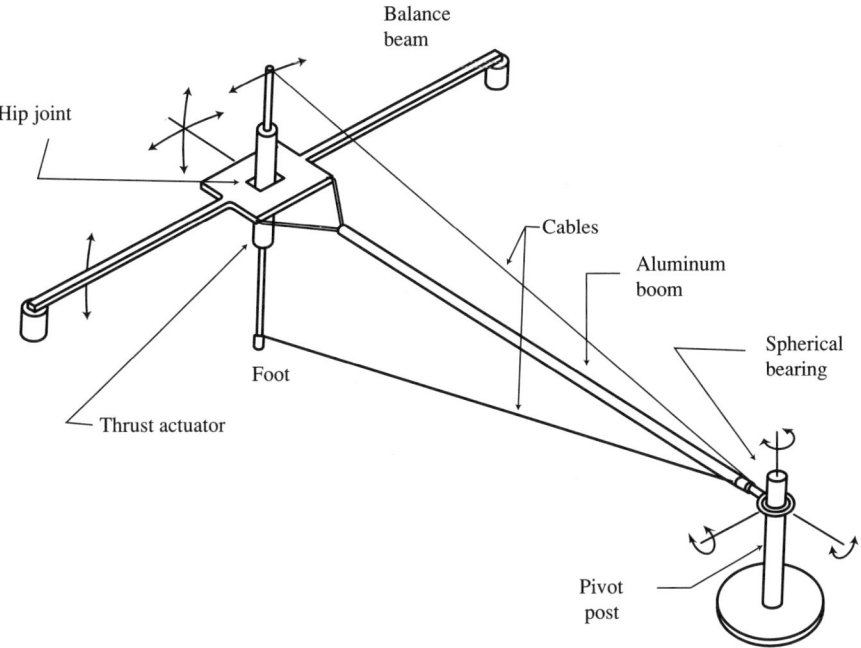

Figure 10.7
Planar one-legged hopping machine. From (Raibert, 1986).

Locomotion can be viewed as a special case of manipulation, since it is a process for moving something around. Many of the processes are similar, and share some of the fundamental mechanics. In particular frictional contact is key.

Legged robots and their gaits are often characterized as either statically stable or dynamically stable. These correspond roughly to our concepts of quasistatic and dynamic manipulation. A typical static analysis is given in figure 10.6. A *support polygon* is defined by the convex hull of all active supports. As long as the center of gravity project to the interior of the support polygon, assuming dynamic forces are negligible, the robot cannot tip over.

Typical analysis of indoor mobile robots might be termed "kinematic locomotion". It is often sufficient to neglect all issues of stability, and assume that wheels do not slip, so that the robot's motion is determined entirely by kinematics. (See, for example, the unicycle and cart examples of section 2.5.)

Dynamic legged locomotion includes hopping, running, or even walking if the dynamic forces are significant. The running and hopping machines built by Raibert and his colleagues illustrate the principles (figure 10.7). They started by building a single-legged running machine that works in two-dimensions, then extended the principles to a variety of machines using 1, 2, or 4 legs, in 2 or 3 dimensions. The machines performed a variety of running gaits (trotting, pacing, bounding), and also performed some gymnastic maneuvers: aerials and flips.

Although Raibert and his collaborators used sophisticated dynamic models in the analysis and design stages, the machines used simple feedback control systems. The vertical hopping cycle is stabilized by injecting a small amount of energy when the robot is on the ground. Body pitch is stabilized by a linear controller that applies hip torque when the foot is on the ground. Forward velocity is controlled by varying the foot placement as the robot strikes the ground.

Anybody familiar with icy pavement knows that frictional contact is important in locomotion, dynamic or otherwise. But there is a real difference between quasistatic locomotion which is stabilized by the geometry of frictional contact alone, versus dynamic locomotion which must stabilize a dynamic cycle.

Subsequent work, described in the bibliographic notes immediately below, has applied some of the lessons of dynamic locomotion to manipulation tasks, and has also explored the stabilization of dynamic locomotion without active servos.

10.4 Bibliographic notes

For a general discussion and survey of dynamic manipulation see (Koditschek, 1993). Raibert (1986) discusses dynamic locomotion and compares locomotion and manipulation.

Quasidynamic manipulation

Erdmann and Mason (1988) explored the automatic analysis and planning with tilting trays, using what we might now call quasidynamic analysis. Yokokohji et al. (1993) were perhaps the first to introduce *quasidynamic* into the lexicon. See (Erdmann, 1998) for more recent work along a similar vein.

Briefly dynamic manipulation

Hove and Slotine (1991) programmed a robot to catch balls. Using stereo vision to predict the ball's path, the arm very quickly matches velocities while the fingers close on the ball.

Dynamic rolling and throwing were also addressed in (Arai and Khatib, 1994; Lynch and Mason, 1999). Impact and dynamic sliding are addressed in (Huang et al., 1995; Huang, 1997).

Juggling

We will define juggling broadly, to include any machine that controls the motions of flying objects by catching, throwing, or hitting. By this definition running is a kind of self-juggling, and ping pong is adversarial juggling, although those topics are considered separately.

The first juggling machine is the Shannon juggler described earlier. Shannon did not describe his dynamic juggler in print, but see (Schaal et al., 1992) for a description of a similar machine that can even juggle five balls.

Sakaguchi et al. (1991); Miyazaki (1993) obtain robot juggling of one or two balls with a single robot hand. The hand makes an elliptical motion, modified by the perceived path of a ball. The hand is a simple funnel, and the balls appear to be very soft. They also describe a robot to play the ball-in-cup game.

THE KODITSCHEK/BUHLER/RIZZI JUGGLERS

Batting means generating a single collision between effector and projectile in order to redirect the projectile. It combines catching and throwing in a single collision. Bühler and Koditschek (1990) describe a machine using a single bar pivoting about its center to bat pucks sliding on an inclined plane. The goal is to achieve a cyclic bouncing of the puck, with stable height, and simultaneously to stabilize its horizontal position measured along the bar. Buhler and Koditschek found a very simple feedback law called a *mirror law* that stabilizes one or two bouncing pucks simultaneously. In essence, the motion of the bar mirrors the motion of the puck. To juggle two pucks, the controller has a mechanism for tending to whichever puck needs immediate attention, as well as a term to keep the two pucks out of phase. Rizzi and Koditschek (1992) extend the principles to three dimensions, to bat two ping pong balls in space.

ATKESON'S MACHINES

Chris Atkeson and his colleagues have developed a number of juggling machines, and use them to explore issues in machine learning. Aboaf et al. (1987) describe a robot that iteratively improves its throwing motion to throw a ball more accurately. Aboaf et al. (1989) describe a system that bats a single ball in three dimensions. It also improves with experience.

Schaal et al. (1992) built a special devil sticking robot. The machine has two "effector sticks" mounted by springy joints to a "torso". A "devil stick" is batted back and forth between the two effector sticks. Each effector stick strikes the devil stick at its center of percussion, halting the effector stick and storing its energy momentarily at a springy joint.

The energy is then transferred back to the devil stick, throwing it to the other effector stick. The problem is simplified by mounting the devil stick so that it has only three degrees of freedom, rather than the usual six.

OTHER JUGGLERS

Miura at the University of Tokyo has constructed robots that can play the Japanese ball-in-cup (kendama) game and spin a top (koma) by throwing it with a string (Miura, 1993).

Ping pong

If we regard batting ping pong balls in the air to be a type of juggling, then playing ping pong might be regarded as adversarial juggling. Russ Andersson built a system to play ping pong at Bell Labs, which is surely one of the most exciting robots ever built (Andersson, 1989).

Robots play a modified form of ping pong: The table dimensions are smaller, the net is higher, and the ball must pass through a square wire frames at each end of the table. The game is thus accessible to current robotic technology. The play is slowed down somewhat, although it is still a challenging dynamic game. Andersson's robot was good enough to beat me, although I believe I would win a rematch.

The robot used multiple cameras to track the ball in three dimensions. A small industrial arm wielded a paddle with an extra-long handle. The robot's planner incorporated a model of ball flight and impact, and used these models to plan a nominal trajectory for the paddle. This nominal trajectory was then refined by iterated simulations, with concurrent adjustment of goals as better estimates of the ball's motion became available.

Dynamic locomotion

Some of the most interesting work in manipulation, robotic juggling in particular, has strong ties to research on dynamic robotic locomotion. The best example may be the running and hopping machines described above (Raibert, 1986).

There are many other machines that demonstrate dynamic walking, or other dynamic tasks. Miura and Shimoyama at the University of Tokyo built a small walking robots with a motion that resembles a person walking with stilts (Miura and Shimoyama, 1984).

McGeer (1990) built a walking machine with a gait resembling a human's, which nas no motors, sensors, or computers. It was designed so that the intrinsic dynamic behavior produces stable walking patterns.

Similarly, there are examples of locomotion work that involve prehension, such as the brachiating robots originally developed by Saito et al. (1994).

Exercises

Exercise 10.1: Provide the details we skipped for figure 10.4. Use Reuleaux's method to identify the acceleration centers for every kinematically feasible contact mode, as in figure 8.8. For each such mode, use moment labeling to construct the corresponding applied wrenches. Add the center of mass, labeled $\pm$, to the labeled regions and take the convex hull. Represent the result as a cone of forces acting through the center of mass.

A Appendix: Infinity

First impressions are sometimes misleading. Points at infinity might first appear to be hopelessly abstract and disconnected from the real world. To the contrary, points at infinity capture the very practical notion of *direction*, and provide pragmatic approaches to a variety of problems.

In this appendix we start with the Euclidean plane $\mathbf{E}^2$ and add some *points at infinity* to obtain the *projective plane* $\mathbf{P}^2$. We then explore some of the basic properties of points at infinity and the projective plane. Homogeneous coordinates provide the easiest construction of points at infinity. Ordinarily we would represent points in the Euclidean plane by two coordinates, (x, y). Homogeneous coordinates introduce a third, apparently redundant, coordinate w, which serves as a scale factor. If we restrict w to be non-zero, we can use homogeneous coordinates to represent the Euclidean plane by the mapping

$$\begin{pmatrix} x \\ y \\ w \end{pmatrix} \mapsto \begin{pmatrix} x/w \\ y/w \end{pmatrix}, w \neq 0 \tag{A.1}$$

It is sometimes convenient to adopt the convention $w = 1$ for points in the Euclidean plane.

Note that we can scale the homogeneous coordinates of a point, without affecting the point:

$$\begin{pmatrix} ax \\ ay \\ aw \end{pmatrix} \mapsto \begin{pmatrix} ax/aw \\ ay/aw \end{pmatrix} = \begin{pmatrix} x/w \\ y/w \end{pmatrix}, a, w \neq 0 \tag{A.2}$$

We now define a *point at infinity* to be a point with homogeneous coordinates $(x, y, 0)$. The *projective plane* refers to the Euclidean plane augmented by the points at infinity.

But where are these points? How do we know that the addition of these points makes geometrical sense? A careful treatment would take us well beyond the purpose of this appendix, but a simple construction yields a number of insights. (See figure A.1). The homogeneous coordinates (x, y, w) form a three-dimensional space. The Euclidean plane is represented by the points with homogeneous coordinates $(x, y, 1)$, which is the plane $w = 1$. Given any point with homogeneous coordinates (x, y, w), with non-zero w, we can scale by $1/w$ and project the point onto the plane. This is a *central projection* of homogeneous coordinate space onto the plane $w = 1$, and gives a nice geometrical picture of the homogeneous coordinate representation of the Euclidean plane. However, the points at infinity map to nowhere under this projection.

To address the homogeneous coordinate representation of points at infinity, we construct the unit sphere in homogeneous coordinate space, and perform a central projection onto the sphere. Each point maps to two antipodal points on the sphere. Now consider the original Euclidean plane $w = 1$. Each such point maps to a single point on the upper hemisphere, and a second point on the lower hemisphere. Consider the motion of the point on the upper hemisphere as a point in the plane "tends toward infinity". The point in the upper hemisphere tends toward the equator, but never quite gets there. If we could manipulate the point on the sphere instead, we could move it to the equator, and into the lower hemisphere. The original Euclidean point would disappear and reappear from the opposite direction.

The central projection onto a sphere is a convenient way of addressing almost any question about the projective plane. As an example, how do we know that the projective plane satisfies Euclid's postulate "two points determine a line"? A point in the projective plane is represented by a pair of antipodal points on the sphere, or, equivalently, a line through the origin of homogeneous coordinate space. Two points in the projective plane would then determine two lines through the origin of homogeneous coordinate space, which would in turn determine a plane through the origin of homogeneous coordinate space. This plane intersects the unit sphere in a great circle, and usually intersects the plane $w = 1$ in a line. So we may think of a line in the projective plane as a plane through the origin of homogeneous coordinate space, or as a great circle on the unit sphere (with antipodal points identified). This construction yields a unique line for any two distinct points in the projective plane.

By similar reasoning, one can show that the points at infinity fall on a single line, the *line at infinity*, which corresponds to the equator of the unit sphere in homogeneous coordinate space. One can also show that parallel lines in the Euclidean plane map to lines intersecting at a point at infinity, and that there are no parallel lines in the projective plane. Since we often think of parallel lines as sharing a common direction, a point at infinity provides a formalism for the common notion of a direction.

Because there are no parallel lines in the projective plane, one might adopt the postulate "two lines determine a point", rather than Euclid's parallel line postulate. This departure from Euclidean geometry yields a duality between points and lines—for every axiom, we can interchange the words "point" and "line" and obtain another axiom. It follows that the same interchange of words, applied to a theorem, will yield another theorem.

Using homogeneous coordinates, one can extend all of these notions to Euclidean three-space. Projective three-space includes a plane of points at infinity. Parallel lines intersect at a point at infinity. Parallel planes intersect at a line at infinity. Projective three space has the same topology as a sphere in Euclidean four-space, with antipodal points identified, which is the same as the set of all lines through the origin of Euclidean four-space.

Infinity

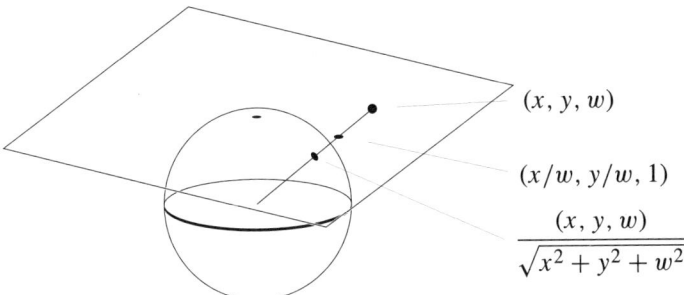

Figure A.1
Central projection of the sphere to a plane. Antipodal points on the sphere map to the plane. Antipodal points on the equator map to the line at infinity.

One further note on nomenclature is in order. Points at infinity are often referred to as *ideal* points, and they form the *ideal line* and the *ideal plane*.

To summarize:

- Points at infinity are mathematically well-defined additions to the Euclidean plane;
- Points at infinity are conveniently represented using the homogeneous coordinates $(x, y, 0)$;
- The points at infinity form a line in the projective plane;
- Parallel lines in the Euclidean plane, when mapped to the projective plane, intersect at a point at infinity;
- The topology of the projective plane is the same as a sphere with antipodal points identified, and the same as the space of all lines through a given point in Euclidean three-space.

References

Aboaf, E. W., C. G. Atkeson, and D. J. Reinkensmeyer (1987). Task-level robot learning: Ball throwing. AI memo 1006, Massachusetts Institute of Technology.

Aboaf, E. W., S. M. Drucker, and C. G. Atkeson (1989). Task-level robot learning: Juggling a tennis ball more accurately. In *IEEE International Conference on Robotics and Automation*, Scottsdale, AZ, pp. 1290–1295.

Alexander, J. C. and J. H. Maddocks (1993). Bounds on the friction-dominated motion of a pushed object. *International Journal of Robotics Research 12*(3), 231–248.

Altmann, S. L. (1989, December). Hamilton, Rodrigues, and the quaternion scandal. *Mathematics Magazine 62*(5), 291–308.

Andersson, R. L. (1989). Understanding and applying a robot ping-pong player's expert controller. In *IEEE International Conference on Robotics and Automation*, Scottsdale, AZ, pp. 1284–1289.

Arai, H. and O. Khatib (1994). Experiments with dynamic skills. In *1994 Japan–USA Symposium on Flexible Automation*, pp. 81–84.

Asada, H. and A. B. By (1985, June). Kinematic analysis of workpart fixturing for flexible assembly with automatically reconfigurable fixtures. *IEEE Journal of Robotics and Automation RA-1*(2), 86–94.

Ball, R. S. (1900). *The Theory of Screws*. Cambridge University Press.

Baraff, D. (1990, April). Determining frictional inconsistency is NP-complete. Department of Computer Science 90-1112, Cornell University.

Baraff, D. (1993). Issues in computing contact forces for non-penetrating rigid bodies. *Algorithmica 10*, 292–352.

Barraquand, J. and J.-C. Latombe (1991). Robot motion planning: A distributed representation approach. *International Journal of Robotics Research 10*, 628–649.

Barraquand, J. and J.-C. Latombe (1993). Nonholonomic multibody mobile robots: Controllability and motion planning in the presence of obstacles. *Algorithmica 10*, 121–155.

Bicchi, A. and V. Kumar (2000). Robotic grasping and contact: A review. In *IEEE International Conference on Robotics and Automation*, pp. 348–353.

Blind, S. J., C. C. McCullough, S. Akella, and J. Ponce (2000). A reconfigurable parts feeder with an array of pins. In *IEEE International Conference on Robotics and Automation*, pp. 147–153.

Boothby, W. M. (1975). *An Introduction to Differentiable Manifolds and Riemannian Geometry*. Academic Press.

Boothroyd, G. (1992). *Assembly Automation and Product Design*. Marcel Dekker.

Bottema, O. and B. Roth (1979). *Theoretical Kinematics*. North-Holland.

Brockett, R. W. (1990). Some mathematical aspects of robotics. *Proceedings of Symposia in Applied Mathematics 41*, 1–19.

Bronowski, J. (1976). *The Ascent of Man*. Boston: Little, Brown.

Brost, R. C. (1988, February). Automatic grasp planning in the presence of uncertainty. *International Journal of Robotics Research 7*(1), 3–17.

Brost, R. C. (1991a, January). *Analysis and Planning of Planar Manipulation Tasks*. Ph. D. thesis, Carnegie Mellon University, School of Computer Science.

Brost, R. C. (1991b). Computing the possible rest configurations of two interacting polygons. In *IEEE International Conference on Robotics and Automation*, Sacramento, CA.

Brost, R. C. and K. Y. Goldberg (1996, February). A complete algorithm for designing planar fixtures using modular components. *IEEE Journal of Robotics and Automation 12*, 31–46.

Brost, R. C. and M. T. Mason (1989, August). Graphical analysis of planar rigid-body dynamics with multiple frictional contacts. In *International Symposium on Robotics Research*, Tokyo, Japan, pp. 293–300. Cambridge, MA: MIT Press.

Bühler, M. and D. E. Koditschek (1990). From stable to chaotic juggling: Theory, simulation, and experiments. In *IEEE International Conference on Robotics and Automation*, Cincinnati, OH, pp. 1976–1981.

Chatterjee, A. and A. Ruina (1998). A new algebraic rigid body collision law based on impulse space considerations. *ASME Journal of Applied Mechanics 65*, 935–950.

Cheng, H. and K. C. Gupta (1989, March). An historical note on finite rotations. *Journal of Applied Mechanics 56*, 139–145.

Collias, N. E. and E. C. Collias (1984). *Nest Building and Bird Behavior*. Princeton University Press.

Crenshaw, J. W. (1994, May). Programmer's toolbox: The hard way. *Embedded Systems Programming*, 11–22.

De Fazio, T. L. and D. E. Whitney (1987). Simplified generation of all mechanical assembly sequences. *IEEE Transactions on Robotics and Automation RA-3*(6), 640–658. errata in RA-6(6), 705–708.

Erdmann, M. (1998). An exploration of nonprehensile two-palm manipulation. *International Journal of Robotics Research 17*(5).

Erdmann, M. A. (1984, August). On motion planning with uncertainty. Master's thesis, Massachusetts Institute of Technology.

Erdmann, M. A. (1994). On a representation of friction in configuration space. *International Journal of Robotics Research 13*(3), 240–271.

Erdmann, M. A. and M. T. Mason (1988, August). An exploration of sensorless manipulation. *IEEE Journal of Robotics and Automation 4*(4), 369–379.

Fujimori, T. (1990). Development of flexible assembly system SMART. In *Proceedings, International Symposium on Industrial Robots*, pp. 75–82.

Gillmor, C. S. (1971). *Coulomb and the Evolution of Physics and Engineering in Eighteenth Century France*. Princeton, New Jersey: Princeton University Press.

Goldberg, K. Y. (1993). Orienting polygonal parts without sensors. *Algorithmica 10*, 201–225.

Goldman, A. J. and A. W. Tucker (1956). Polyhedral convex cones. In H. W. Kuhn and A. W. Tucker (Eds.), *Linear Inequalities and Related Systems*, pp. 19–39. Princeton, NJ: Princeton University Press. Volume 38, Annals of Mathematics Studies.

Goyal, S. (1989). *Planar Sliding of a Rigid Body With Dry Friction: Limit Surfaces and Dynamics of Motion*. Ph. D. thesis, Cornell University, Dept. of Mechanical Engineering.

Goyal, S., A. Ruina, and J. Papadopoulos (1991). Planar sliding with dry friction. Part 1. Limit surface and moment function. *Wear 143*, 307–330.

Grossman, D. D. and M. W. Blasgen (1975, September). Orienting mechanical parts by computer-controlled manipulator. *IEEE Transactions on Systems, Man, and Cybernetics*.

Guibas, L. J., L. Ramshaw, and J. Stolfi (1983, November). The kinetic framework for computational geometry. In *Proc. the 24th Annual Symposium on Foundations of Computer Science (FOCS)*, Tucson, Arizona, pp. 100–111. IEEE.

Halperin, D., L. Kavraki, and J.-C. Latombe (1997). *Robotics*, Chapter 41, pp. 755–778. CRC Press.

Hanafusa, H. and H. Asada (1977). Stable prehension by a robot hand with elastic fingers. In *Proceedings of the Seventh International Symposium on Industrial Robots*, pp. 361–368.

Hartenberg, R. S. and J. Denavit (1964). *Kinematic Synthesis of Linkages*. McGraw-Hill.

Hilbert, D. and S. Cohn-Vossen (1952). *Geometry and the Imagination*. Chelsea.

Hirai, S. and H. Asada (1993, October). Kinematics and statics of manipulation using the theory of polyhedral convex cones. *International Journal of Robotics Research 12*(5), 434–447.

Homem de Mello, L. S. and A. C. Sanderson (1990). AND/OR graph representation of assembly plans. *IEEE Transactions on Robotics and Automation 6*(2), 188–199.

Hove, B. and J. Slotine (1991, June). Experiments in robotic catching. In *Proceedings of the 1991 American Control Conference*, pp. 380–385.

Howe, R. D. and M. R. Cutkosky (1996, December). Practical force-motion models for sliding manipulation. *International Journal of Robotics Research 15*(6), 557–572.

Huang, W., E. P. Krotkov, and M. T. Mason (1995). Impulsive manipulation. In *IEEE International Conference on Robotics and Automation*, pp. 120–125.

Huang, W. H. (1997, August). *Impulsive Manipulation*. Ph. D. thesis, Carnegie Mellon University. CMU-RI-TR-97-29.

References

Hunt, K. H. (1978). *Kinematic Geometry of Mechanisms*. Oxford University Press.

Kane, T. R. and D. A. Levinson (1978, December). Successive finite rotations. *Journal of Applied Mechanics 45*.

Kerr, J. and B. Roth (1986). Analysis of multifingered hands. *International Journal of Robotics Research 4*(4), 3–17.

Khatib, O. (1980). *Commande Dynamique dans l'Espace Opérationnel des Robots Manipulateurs en Présence d'Obstacles*. Ph. D. thesis, Ecole Nationale Supérieure de l'Aéronautique et de l'Espace, Toulouse.

Khatib, O. (1986). Real-time obstacle avoidance for manipulators and mobile robots. *International Journal of Robotics Research 5*(1), 90–98.

Koditschek, D. E. (1993). Dynamically dexterous robots. In M. W. Spong, F. L. Lewis, and C. T. Abdallah (Eds.), *Robot Control: Dynamics, Motion Planning and Analysis*. New York: IEEE Press.

Korn, G. A. and T. M. Korn (1968). *Mathematical Handbook for Scientists and Engineers* (second ed.). New York: McGraw-Hill.

Lakshminarayana, K. (1978). Mechanics of form closure. ASME Rep. 78-DET-32, 1978.

Latombe, J.-C. (1991). A fast path planner for a car-like indoor mobile robot. In *National Conference on Artificial Intelligence*, pp. 659–665.

Li, Z. and J. Canny (1990). Motion of two rigid bodies with rolling constraint. *IEEE Transactions on Robotics and Automation 6*, 62–72.

Lin, Q. and J. W. Burdick (2000, June). Objective and frame-invarient kinematic metric functions for rigid bodies. *International Journal of Robotics Research 19*(6), 612–625.

Lozano-Pérez, T. and M. A. Wesley (1979). An algorithm for planning collision-free paths among polyhedral obstacles. *Communications of the ACM 22*, 560–570.

Lynch, K. M. and M. T. Mason (1996, December). Stable pushing: Mechanics, controllability, and planning. *International Journal of Robotics Research 15*(6), 533–556.

Lynch, K. M. and M. T. Mason (1999, January). Dynamic nonprehensile manipulation: Controllability, planning and experiments. *International Journal of Robotics Research 18*(1), 64–92.

MacMillan, W. D. (1936). *Dynamics of Rigid Bodies*. Dover, New York.

Mason, M. T. (1986, Fall). Mechanics and planning of manipulator pushing operations. *International Journal of Robotics Research 5*(3), 53–71.

Mason, M. T. (1991, November). Two graphical methods for planar contact problems. In *IEEE/RSJ International Conference on Intelligent Robots and Systems*, Osaka, Japan, pp. 443–448.

Mason, M. T. and K. M. Lynch (1993). Dynamic manipulation. In *IEEE/RSJ International Conference on Intelligent Robots and Systems*, Yokohama, Japan, pp. 152–159.

Mason, M. T. and J. K. Salisbury, Jr. (1985). *Robot Hands and the Mechanics of Manipulation*. The MIT Press.

McCarthy, J. M. (1990). *Introduction to Theoretical Kinematics*. MIT Press.

McGeer, T. (1990). Passive dynamic walking. *International Journal of Robotics Research 9*(2), 62–82.

McGhee, R. B. and A. A. Frank (1968). On the stability properties of quadruped creeping gaits. *Mathematical Biosciences 3*, 331–351.

Mishra, B., J. T. Schwartz, and M. Sharir (1987). On the existence and synthesis of multifinger positive grips. *Algorithmica 2*(4), 541–558.

Miura, H. (1993). Private communication.

Miura, H. and I. Shimoyama (1984). Dynamic walk of a biped. *International Journal of Robotics Research 3*(2), 60–74.

Miyazaki, F. (1993). Motion Planning and Control for a Robot Performer (videotape).

Montana, D. J. (1988, June). The kinematics of contact and grasp. *International Journal of Robotics Research 7*(3), 17–32.

Murray, R. M., Z. Li, and S. S. Sastry (1994). *A Mathematical Introduction to Robotic Manipulation*. CRC Press.

Napier, J. (1993). *Hands*. Princeton University Press.

Nguyen, V.-D. (1988). Constructing force-closure grasps. *International Journal of Robotics Research 7*(3).

Ohwovoriole, M. S. (1980). *An Extension of Screw Theory and Its Application to the Automation of Industrial Assembly*. Ph. D. thesis, Stanford University.

Okamura, A. M., N. Smaby, and M. R. Cutkosky (2000). An overview of dexterous manipulation. In *IEEE International Conference on Robotics and Automation*, pp. 255–262.

Overton, M. L. (1983). A quadratically convergent method for minimizing a sum of Euclidean norms. *Math. Programming 27*, 34–63.

Painlevé, P. (1895). Sur les lois du frottement de glissement. *Comptes Rendus de l'Académie des Sciences 121*, 112–115.

Pang, J. and J. Trinkle (1996). Complementarity formulations and existence of solutions of dynamic multi-rigid-body contact problems with Coulomb friction. *Mathematical Programming 73*, 199–226.

Paul, B. (1979). *Kinematics and Dynamics of Planar Machinery*. New Jersey: Prentice-Hall.

Peshkin, M. A. and A. C. Sanderson (1988a, December). The motion of a pushed, sliding workpiece. *IEEE Journal of Robotics and Automation 4*(6), 569–598.

Peshkin, M. A. and A. C. Sanderson (1988b, October). Planning robotic manipulation strategies for workpieces that slide. *IEEE Journal of Robotics and Automation 4*(5), 524–531.

Peshkin, M. A. and A. C. Sanderson (1989, February). Minimization of energy in quasi-static manipulation. *IEEE Transactions on Robotics and Automation 5*(1), 53–60.

Prescott, J. (1923). *Mechanics of Particles and Rigid Bodies*. Longmans, Green, and Co., London.

Raibert, M. H. (1986). *Legged Robots That Balance*. Cambridge: MIT Press.

Rajan, V. T., R. Burridge, and J. T. Schwartz (1987, March). Dynamics of a rigid body in frictional contact with rigid walls. In *IEEE International Conference on Robotics and Automation*, Raleigh, North Carolina, pp. 671–677.

Reuleaux, F. (1876). *The Kinematics of Machinery*. MacMillan. Reprinted by Dover, 1963.

Rimon, E. and J. W. Burdick (1995, June). New bounds on the number of frictionless fingers required to immobilize planar objects. *Journal of Robotic Systems 12*(6), 433–451.

Rizzi, A. A. and D. E. Koditschek (1992). Progress in spatial robot juggling. In *IEEE International Conference on Robotics and Automation*, Nice, France, pp. 775–780.

Roth, B. (1984). Screws, motors, and wrenches that cannot be bought in a hardware store. In M. Brady and R. Paul (Eds.), *Robotics Research: The First International Symposium*, pp. 679–693. MIT Press.

Routh, E. J. (1913). *Dynamics of a System of Rigid Bodies*. MacMillan and Co.

Saito, F., T. Fukuda, and F. Arai (1994, February). Swing and locomotion control for a two-link brachiation robot. *IEEE Control Systems Magazine 14*(1), 5–12.

Sakaguchi, T., Y. Masutani, and F. Miyazaki (1991). A study on juggling task. In *IEEE/RSJ International Conference on Intelligent Robots and Systems*, Osaka, Japan, pp. 1418–1423.

Salamin, E. (1979). Application of quaternions to computation with rotations. Internal Working Paper, Stanford Artificial Intelligence Lab.

Salisbury, Jr., J. K. (1982). *Kinematic and Force Analysis of Articulated Hands*. Ph. D. thesis, Stanford University.

Savage-Rumbaugh, S. and R. Lewin (1994). *Kanzi: The Ape at the Brink of the Human Mind*. John Wiley and Sons.

Schaal, S., C. G. Atkeson, and S. Botros (1992). What should be learned? In *Seventh Yale Workshop on Adaptive and Learning Systems*, pp. 199–204.

Simunovic, S. N. (1975, September 22–24). Force information in assembly processes. In *Proceedings, 5th International Symposium on Industrial Robots*.

Stewart, D. E. (1998). Convergence of a time-stepping scheme for rigid-body dynamics and resolution of Painlevé's problem. *Arch. Rational Mech. Anal. 145*, 215–260.

References

Stewart, D. E. (2000). Rigid-body dynamics with friction and impact. *SIAM Review 42*(1), 3–39.

Stolfi, J. (1988, May). *Primitives for Computational Geometry*. Ph. D. thesis, Stanford Univ., Department of Computer Science.

Symon, K. R. (1971). *Mechanics*. Addison-Wesley.

Trinkle, J. C. (1992, October). On the stability and instantaneous velocity of grasped frictionless objects. *IEEE Transactions on Robotics and Automation 8*(5), 560–572.

Trinkle, J. C., J.-S. Pang, S. Sudarsky, and G. Lo (1997). On dynamic multi-rigid-body contact problems with Coulomb friction. *Z. angew. Math. Mech. 77*(4), 267–279.

Truesdell, C. (1968). *Essays in the History of Mechanics*. New York: Springer-Verlag.

Wang, Y. and M. T. Mason (1992, September). Two-dimensional rigid-body collisions with friction. *ASME Journal of Applied Mechanics 59*, 635–641.

Wilson, F. R. (1998). *The Hand*. New York: Pantheon.

Yokokohji, Y., Y. Yu, N. Nakasu, and T. Yoshikawa (1993). Quasi-dynamic manipulation of constrained object by robot fingers in assembly tasks. In *Proceedings of 1993 IEEE/RSJ International Conference on Intelligent Robots And Systems*, pp. 144–151.

Author Index

Aboaf, 233
Alexander, 174
Altmann, 51, 72
Amontons, 123, 139
Andersson, 234
Arai, 232, 234
Aristotle, 121
Asada, 117, 173
Atkeson, 233

Ball, 37
Baraff, 199, 207
Barraquand, 83, 85, 88
Bicchi, 88, 173, 174
Blasgen, 174
Blind, 174
Boothroyd, 8
Botros, 233
Bottema, 36
Brockett, 36
Bronowski, 8
Brost, 116, 117, 156, 173, 174, 177, 178
Buhler, 233
Burdick, 36, 37
By, 173

Canny, 88
Chatterjee, 223
Chebyshev, 39
Cheng, 51
Cohn-Vossen, 36
Collias, 8
Coulomb, 123, 139
Crenshaw, 72
Cutkosky, 88, 139, 140, 174

Da Vinci, Leonardo, 123, 139
De Fazio, 174
Denavit, 36
Drucker, 233

Erdmann, 117, 139, 174, 207, 232
Euclid, 238

Fujimori, 8
Fukuda, 234

Gauss, 51
Gillmor, 139
Goldberg, 163, 165, 167, 173, 174
Goldman, 117
Goyal, 134
Grossman, 174

Guibas, 117
Gupta, 51

Halperin, 174
Hamilton, 50
Hanafusa, 173
Hartenberg, 36
Hilbert, 36
Hirai, 117
Homem de Mello, 174
Hove, 232
Howe, 139, 140, 174
Huang, 232
Hunt, 37, 117

Kane, 72
Kavraki, 174
Khatib, 88, 232
Koditschek, 233
Korn, 72
Krotkov, 232
Kumar, 88, 173, 174

Lakshminarayana, 174
Latombe, 83, 85, 88, 174
Levinson, 72
Lewin, 8
Li, 36, 72, 88
Lin, 36
Liouville, 229
Lozano-Pérez, 88
Lynch, 8, 174, 232

MacMillan, 139
Maddocks, 174
Mason, 8, 117, 139, 156, 174, 177, 232
Masutani, 233
McCarthy, 36
McGeer, 234
Mishra, 173
Miura, 234
Miyazaki, 233
Montana, 88
Murray, 36, 72, 88

Nakasu, 232
Napier, 8
Newton, 121
Nguyen, 117, 120, 171, 174

Ohwovoriole, 72, 117
Okamura, 88
Overton, 174

Painlevé, 207
Pang, 117
Papadopoulos, 134, 139
Paul, 37
Peshkin, 156, 174
Prescott, 139

Raibert, 231
Ramshaw, 117
Reinkensmeyer, 233
Reuleaux, 36, 40, 174, 179
Rimon, 37
Rizzi, 233
Roth, 36, 72, 117
Routh, 223
Ruina, 134, 223

Saito, 234
Sakaguchi, 233
Salamin, 72
Salisbury, 86–88, 91, 92
Sanderson, 174
Sastry, 36, 72, 88
Savage-Rumbaugh, 8
Schaal, 233
Schwartz, 173
Shannon, 229
Sharir, 173
Shimoyama, 234
Simunovic, 139, 171, 174
Slotine, 232
Smaby, 88
Stewart, 207, 223
Stolfi, 117
Symon, 117, 207

Trinkle, 117, 174, 207
Truesdell, 139
Tucker, 117

Wang, 223
Watt, 39
Wesley, 88
Whitney, 171, 174
Wilson, 8

Yokokohji, 232
Yoshikawa, 232
Yu, 232

Subject Index

acceleration center, 117, 204
allen wrench, 206, 225–227
ambient space, 11
and/or graph, 169, 174, 180
angle between two lines, 65
angle of repose, 124
angular inertia, 191, 201
angular inertia matrix, 187–195, 208
    parallel axis theorem, 193, 208
    symmetry theorems, 192, 208
angular inertia tensor,
    See angular inertia matrix
angular momentum,
    See moment, of momentum
angular velocity vector, 47, 67
APPROACH, 78
arm, 82
assembly, 3–8, 78, 128, 168–174
    peg in hole, 171–173, 180
    sequence, 168, 180
    two-handed, 169
astroid, 19
axis-angle, 42–43, 46–47, 55–56
axodes, moving and fixed, 24

ball catching, 232
ball-in-cup, 233, 234
Best First Planner, 83–84, 90
BFP,
    See Best First Planner
block in corner, 209
block on table, 128
book throwing, 208
bound vectors, 94
bowl feeder, 162
Brost triangle, 178

cart, 27–28, 231
Cayley-Klein parameters, 74
center of friction, 133, 149, 156, 164, 165, 177–179
center of mass, 184
central projection, 106–109, 112, 113, 237
centrodes, 17–19, 38, 39
change of reference, 95
Chasles's theorem, 22, 24, 66, 98
CLOSE, 77
coefficient of friction, 122, 129, 176, 180, 203
    measurement of, 176
coefficient of restitution, 212
collision avoidance, 78
complex numbers, 50, 52
compliance
    center of, 173
    passive, 173
    remote center of, 173
compliant motion, 77
cones, fixed and moving, 22
cones, moving and fixed, 197
configuration, 11
configuration space, 11, 28, 79
    metric on, 11, 36
configuration space transform, 79–84, 90
connectivity, 91,
    See mechanism, connectivity of
constant diameter, 179
constraint, 18, 25–34, 37, 70, 98
    bilateral, 26, 27, 68, 69
        analysis of, 18–19
    holonomic, 27, 40
    nonholonomic, 25, 27–32, 36
    Pfaffian, 29, 40
    representation of, 68–72
    rheonomic, 25–27
    scleronomic, 27
    unilateral, 25–27, 68, 69
        analysis of, 33–34, 40
contact, 25, 86, 102
    taxonomy of, 86, 87, 91
contact modes, 124, 127, 149, 198, 200, 205, 206, 215, 220
contact normals, 33, 69, 102, 105, 163, 164
contact screw, 69
contact screws, 69, 105, 143
contact wrenches, 102, 105
convex hull, 81, 89, 100, 107–109, 118, 146, 172, 177, 178, 231
conveyor, 147
cost function, 85, 91
couch, 178
Coulomb's law, 121–127, 130, 133, 135, 157, 160, 163, 198, 201, 203, 218, 221
couple, 95, 96
cross-product matrix, 46, 218

degenerate mass distribution, 208
degrees of freedom, 12, 27, 34, 37, 72
DEPROACH, 78
design for assembly, 5, 78, 170
devil sticking, 233
dexterous manipulation,
    See manipulation, in the hand
diameter function, 179
dimensional synthesis, 87
dinner plate, 130, 131
dipods, 156
directed lines, 64
direction vector, 61
displacements, 12

decomposition, 14–15
   group, 14
   planar, 17
      non-commutativity of, 37
   representation, 58–72,
      See homogeneous coordinates; screw coordinates
distance between two lines, 65
distribution, 30
   involutive, 31
   involutive closure of, 31
   regular, 31
drawers, 171
duality of motion and force, 98
dynamic load, 203
dynamics, 181–209
   force dual method, 203–207
   moment labeling, 203–207
   multiple frictional contacts, 205–207
   of a particle, 181–184
   of a system of particles, 184–186
   of rigid bodies, 186–197
   single frictional contact, 197–205

equilibrium, 140, 144, 203
equivalent systems of forces, 94, 96–99
Euclidean plane, 237
Euler angles, 47–51, 72, 73
Euler parameters, 51,
   See quaternions
Euler's equations, 188, 195, 208
Euler's theorem, 20, 22, 36, 42

fence, 147, 148
field, 52
fixed and moving planes, 13, 18
fixtures, 103, 128, 130, 143–147, 169, 173
flexible manufacturing, 162
foliation, 28
force, 93–121
   external, 184
   internal, 184
   total, 184
force closure, 103–104, 116–117, 119, 120, 129, 130, 143–145, 170–172, 174–175
   existence of, 145
   synthesis, 145–147
force dual, 112–120, 127, 128, 139, 143, 155, 171–173, 203–207, 209
form closure, 143–145, 169, 170, 174
   first order, 116–117, 144–145
four-bar linkage, 18, 36
free vector, 94
friction, 121–141
   dynamic, 122

   static, 121
friction angle, 124
friction cone, 124–127, 149, 163, 201, 216, 220
frictional load, 134
Frobenius's theorem, 31, 40
furniture, 147

getting dressed, 180
gimbal, 73
gimbal coordinates, 73
GRASP, 145
grasp, 3, 5, 87, 103–104, 121, 128, 143–147, 168, 173, 175
   planning, 77, 86–88, 169
gripper, 78, 169
Grübler's formula, 36, 86–88
gymnastics, 232

half-space, 101
Hamiltonian system, 229
herpolhode, 196
higher pair, 34
hinge, 147, 148
homogeneous coordinates, 58–60, 72, 106, 237

IC,
   See instantaneous center
ideal points, line, and plane,
   See points at infinity, etc.
impact, 121, 127, 202, 211–223, 225
   elastic, 212
   frictional, 214–216
   Newtonian restitution, 212, 221
   plastic, 212
   Poisson restitution, 212–213, 221
   rigid body, 217–221
   single particle, 211–216
   sliding rod, 219–221
   tangential, 221
impulse, 182, 211
impulse space, 215, 217, 220
impulse-momentum equations, 211, 217, 218, 221
impulsive moment, 211
inconsistency
   frictional, 198, 200–202, 211
indeterminacy
   frictional, 198, 202–203, 209
inertia ellipsoid, 192, 196, 208
infinity, 17, 237–239
instantaneous center, 17, 109, 117, 134, 153, 154, 156, 157, 161
   for pushing, 149–151
instantaneous screw axis, 24

Subject Index

invariable plane, 196
involutive closure, 85

jamming, 170–173, 180
joint, 34
juggling
    Shannon juggler, 229

kendama, 234
kinematic constraint, 40
kinematics, 11–40
    planar, 15–19
    spatial, 22–25
    spherical, 20–22
kinetic energy, 182, 189, 196, 225
koma, 234

leaf of foliation, 28
Lie bracket motions, 31, 85
Lie brackets, 28, 31–32
limit curve, 135
limit surface, 130, 134–140, 151, 174
line at infinity, 62, 238
line of action, 94, 110, 114, 117–119
line of force, 99, 148–152
line of motion, 148–152
line of pushing, 148–152
line vectors, 94
line, representation of, 61
linear algebra, 52
linear span, 99
link, 34
linkage, 18, 36, 86,
    See mechanism
    Chebyshev, 39
    slider-crank, 38
    Watt's, 39
Liouville's theorem, 229
locomotion, 230
    dynamically stable, 231
    kinematic, 231
    statically stable, 231
lower pairs, 34, 35

manhole covers, 179
manipulation, 1–8
    animal, 8
    briefly dynamic, 229
    continuously dynamic, 230–232
    dynamic, 8, 181, 225–235
    human, 1–3, 5–8
    in the hand, 77, 86–88
    industrial, 3–7, 77–78
    kinematic, 7, 77–92, 225
    meaning of, ix, 1, 121
    pick and place, 5, 77–78
    quasidynamic, 225–227
    quasistatic, 8, 143–181, 225, 231
    static, 7, 181, 225
    taxonomy, 7–8
manufacturing
    automated, 168
    flexible, 162
mass, 181
maximum power inequality, 135, 174
mechanisms, 34–36
    connectivity of, 34–36, 88
    mobility of, 34–36
minimum power principle, 151, 174
Minkowski sum and difference, 81, 138
mirror law, 233
mobile manipulation, 153, 177
mobile robots, 38, 88, 89, 153, 161, 231
mobility, 88,
    See mechanism, mobility of
moment
    of a line, 63, 64, 183
    of force, 93, 183–185, 191
    of inertia, 190
    of momentum, 183–186, 196
moment labeling, 110–112, 117–119, 127, 128, 139,
    140, 143, 171–173, 203–207, 209
moment vector, 61
momentum, 181, 182, 184, 225
mouse, 178
MOVETO, 77, 78
moving and fixed planes, 13, 18

Newton's laws, 181–186, 197, 199, 211
Newton-Euler equations, 181, 195
NHP,
    See NonHolonomic Planner
non-Euclidean geometry, 238
nonholonomic
    constraints,
    See constraint, nonholonomic
    systems, 77, 84–86, 161–162
NonHolonomic Planner, 85, 91, 153
number synthesis, 87

OPEN, 77
oriented plane, 105–118

pair
    cylindrical, 174
    helical, 35

higher, 34
lower, 34, 35
planar, 35, 174
prismatic, 35
revolute, 35
spherical, 35
parallel axis theorem, 193
parallel line postulate, 238
parallel lines, 238
particle, 93
parts feeding and parts orienting, 3, 78, 147, 148, 162–168, 174, 178, 179, 225–227
path, 18
path planning, 77–86, 90, 161–162, 169
Pauli spin matrices, 74
pencils, 143
perimeter, 115
phase space, 229
pick, 176
pick and place,
   See manipulation, pick and place
ping pong, 234
pipe clamp, 125–126, 139
pitch, 24, 65, 66, 68, 97, 98
Plücker coordinates, 61–65, 67, 69, 70, 98, 183
planning
   sequence of pushes, 166–168
plastic impact, 214
Poinsot's construction, 196, 197
Poinsot's theorem, 97–98
polhodes, 196, 197
polyhedral convex cones, 68, 99–101, 103–105, 109, 118, 160
positive linear span, 99, 103–105
potential field, 83
power, 98, 182
pressure distribution, 131–133, 137, 154, 156
principal axes, 188, 190, 196
principal moments of inertia, 190
priority queue, 83
products of inertia, 190
projective plane, 17, 107, 237–239
projective three-space, 59, 237–239
puck, 233
push function, 163–168
pushing, 38, 147–168, 174, 176, 178, 179
   bisector bound, 156–161, 177
   Peshkin's bound, 154–161
   square, 163, 164, 180
   stable, 153–162, 177
   vertical strip bound, 157–161
   with known support distribution, 174

quasistatic assumption, 163
quaternions, 47, 50–58, 72, 74, 75

radius function, 163–164, 178–180
radius of gyration, 191
ray, 99
reciprocal product, 64, 69, 70, 98
refrigerator, 38, 91
resultant, 94–96, 110, 116
Reuleaux's method, 33–34, 40, 68, 69, 105–110, 112, 117, 143, 206, 209
rigid body, 12
Rodrigues's formula, 42–43, 46, 51, 55
rotating body, 195–197
rotation axis, 20, 42, 56, 69, 98
rotation center, 17, 33–34
rotation matrices, 43–47, 49–50, 56–57, 74
rotation pole,
   See rotation center
rotations, 13, 20–22
   composition of, 45, 48, 55
   differential, 47
   non-commutativity of, 15, 37, 41, 73
   random, 58, 75
   representation, 41–58,
   See axis-angle; Euler angles; quaternions; rotation matrices
   topology of, 41
Routh's construction, 215, 217, 221, 223
rubber-soled shoes, 221
running, 230–232

screw, 23, 24, 65, 97
   contrary, 69
   reciprocal, 69, 98, 105
   repelling, 69, 105
screw axis, 24
screw coordinates, 60–72, 98, 103
   for planar motion, 71
   of a planar wrench, 98
   of a wrench, 97–98
screw displacement, 22, 23
screw theory, 37
SE(2), **SE**(3), 14, 79
SE(4), 72
SE(n),
   See displacements,
   See special Euclidean group
Shannon juggler, 230
shoes, socks, shorts, and long pants, 180
slider-crank, 38
sliding, 130–139
   force and moment of, 131–134
   indeterminacy of, 130–132, 147

Subject Index

sliding barbell, 137, 138
sliding block, 121, 123, 124
sliding rod, 126–127, 198–203, 209, 211, 219–221
SO(3), 14
SO(n),
　See rotations,
　See special orthogonal group
Sony APOS system, 162, 168, 170
Sony SMART system, 3–6, 8, 78
special Euclidean group, 14, 79,
　See displacements
special orthogonal group, 14,
　See rotations
　　metric, 57–58
　　topology, 57–58
　　uniform distribution, 58
　　uniform distribution on, 75
spherical joints, 86
squeezing, 179
stability, 144
stable placement, 77
static equilibrium, 128–130, 144
static indeterminacy, 129
statics, 93–121
Steinitz's theorem, 146
supplementary cones, 101, 105, 117, 118
support line, 164
support polygon, 231
symmetry, 164, 168
system, 11

table, 143
table tennis,
　See ping pong
tangent space, 29
tip line, 155–156
tolerances, 170
top spinning, 234
torque,
　See moment, of force
trajectory, 18
transform matrix, 59–60, 72,
　See homogeneous coordinates
translation, 13
tray tilting, 225–227
tumbling, 208
tumbling body, 186
twist, 24, 96, 104
　differential, 67–68, 70
　　representation of, 67–68
　representation of, 66–67
type synthesis, 87

uncertainty, ix, 147, 165–166, 229
　conservative approximation of, 166
　in shape, 170
　possibilistic model of, 166
　probabilistic model of, 166
unicycle, 28–32, 85, 86, 231

vector field, 29–32
velocity center,
　See instantaneous center
velocity dual, 117
velocity pole,
　See instantaneous center
virtual product,
　See reciprocal product
visibility graph, 79, 83

wedging, 170–173
work, 182
wrench, 96, 97

zigzag locus, 115, 119, 140, 145, 176